From **Bricolage** *to* **Métissage**

(RE)THINKING ENVIRONMENTAL EDUCATION

Constance Russell and Justin Dillon
General Editors

Vol. 8

The [Re]Thinking Environmental Education series
is part of the Peter Lang Education list.
Every volume is peer reviewed and meets
the highest quality standards for content and production.

PETER LANG
New York • Bern • Frankfurt • Berlin
Brussels • Vienna • Oxford • Warsaw

Gregory Lowan-Trudeau

From Bricolage *to* Métissage

Rethinking Intercultural Approaches to Indigenous Environmental Education and Research

PETER LANG
New York • Bern • Frankfurt • Berlin
Brussels • Vienna • Oxford • Warsaw

Library of Congress Cataloging-in-Publication Data

Lowan-Trudeau, Gregory.
From bricolage to métissage: Rethinking intercultural approaches to indigenous environmental education and research / Gregory Lowan-Trudeau.
pages cm. — ([Re]thinking environmental education; v. 8)
Includes bibliographical references and index.
1. Indigenous peoples—Ecology. 2. Traditional ecological knowledge—Canada.
3. Traditional ecological knowledge—British Columbia. 4. Ecology—Research.
5. Environmental education. 6. Environmental sciences—Research.
7. Biodiversity conservation—Law and legislation—Case studies. I. Title.
GF50.L69 363.70071—dc23 2015003549
ISBN 978-1-4331-2236-1 (hardcover)
ISBN 978-1-4331-2235-4 (paperback)
ISBN 978-1-4539-1527-1 (e-book)
ISSN 1949-0747

Bibliographic information published by **Die Deutsche Nationalbibliothek**.
Die Deutsche Nationalbibliothek lists this publication in the "Deutsche Nationalbibliografie"; detailed bibliographic data are available on the Internet at http://dnb.d-nb.de/.

Cover art by Rachel Mishenene

29 Broadway, 18th floor, New York, NY 10006
www.peterlang.com

DEDICATION

For Cedar 湖児 and Miho

CONTENTS

	Acknowledgements	ix
	Author's Note	xi
	Related Publications	xiii
	List of Tables & Figures	xv
	Prelude: A Tale of Two First Nations	1
Chapter 1.	Introduction	5
Chapter 2.	Methodological Métissage	17
Chapter 3.	Articulating a *Métis* Worldview: Exploring the Third Space	31
Chapter 4.	In Search of Common Ground: To Blend or Not to Blend?	55
Chapter 5.	Environmental Educators' Perspectives	81
Chapter 6.	Three-Eyed Seeing?	101

Chapter 7. Implications for Environmental Education in Canada and Beyond 111
Chapter 8. Final Thoughts & Future Directions 125

References 133
Index 147

ACKNOWLEDGEMENTS

First and foremost I would like to acknowledge the ongoing support and encouragement of my family. I am grateful to you all for your encouragement and guidance. Thank you. Merci. Wela'lin.

I would also like to recognize and extend deep gratitude to the participants in both studies for their generous insights.

The staff and faculty of the University of Calgary's Werklund School of Education also deserve great appreciation, most especially my doctoral supervisor, Dr. Gail Jardine, for her unwavering commitment and dedication to facilitating my research journey. Many thanks also to my doctoral supervisory committee that involved, at different stages, Dr. Ann Sherman, the late Dr. Jeff Jacob, Dr. Cecille DePass, and Dr. Sylvie Roy. Thank you all for your support, encouragement, and adaptability. Thanks also to the external examiners of my doctoral dissertation, Dr. Cynthia Chambers and Dr. Mishka Lysack.

Warm thanks also to the staff and faculty of Lakehead University's Faculty of Education for providing me with a physical, intellectual, and existential home-away-from-home during the last two years of my doctoral studies. Special thanks especially to Dr. Connie Russell and Dr. Bob Jickling for your ongoing insight, encouragement, and friendship. Thanks also to Dr. Russell and Dr. Justin Dillon for inviting me to contribute to this series and for your excellent editorial support.

I would also like to acknowledge the financial support provided by the Social Sciences and Humanities Research Council of Canada, the Killam Foundation, the University of Calgary, and the University of Northern British Columbia.

Finally, I would like to extend warm appreciation to Rachel Mishenene for contributing the beautiful painting found on the cover of this book. Chi Meegwetch.

AUTHOR'S NOTE

100% of the author's profits from this book will be donated to Indigenous and environmental education initiatives.

RELATED PUBLICATIONS

Modified excerpts from the following publications are gratefully reproduced here with permission:

Lowan-Trudeau, G. (under review). Three-eyed seeing? Considering Indigenous ecological knowledge in culturally complex contexts. *Alberta Science Education Journal*.

Lowan-Trudeau, G. (in press). Indigenous environmental education in North America and beyond. In C. Russell, J. Dillon & M. Breunig (Eds.). *Environmental Education Reader*. New York: Peter Lang.

Lowan-Trudeau, G. (2013). Indigenous environmental education research in North America: A brief review. In R. Stevenson, M. Brody, J. Dillon & A.E.J. Wals (Eds.), *International handbook of research on environmental education* (pp. 404–408). New York: Routledge & The American Educational Research Association.

Lowan-Trudeau, G. (2014). Considering ecological métissage: To blend or not to blend? *Journal of Experiential Education*, *37*(4), 351–366.

Lowan-Trudeau, G. (2013). Navigating the wilderness between us: Exploring ecological métissage as an emerging vision for environmental education in Canada [Thesis Synopsis]. *Environmental Education Research*, *19*(2), 253–254.

Lowan-Trudeau, G. (2012). Methodological métissage: An interpretive Indigenous approach to environmental education research. *Canadian Journal of Environmental Education*, *17*, 113–130.

Lowan, G. (2012). Expanding the conversation: Further explorations into Indigenous environmental science education theory, research, and practice. *Cultural Studies in Science Education*, *7*, 71–81.

Lowan, G. (2011). Adrift in our national consciousness: Meditations on the canoe. *Pathways: The Ontario Journal of Outdoor Education*, *23*(4), 25–29.

Lowan, G. (2011). Ecological métissage: Exploring the third space in Canadian outdoor and environmental education. *Pathways: The Ontario Journal of Outdoor Education*, *23*(2), 10–15.

LIST OF TABLES & FIGURES

Table 1.	Methodological Métissage (Lowan-Trudeau, 2012a; 2012b)	21
Figure 1.	Methodological Métissage (Lowan-Trudeau, 2012a, 2012b)	21
Figure 2.	The Third Space (Lowan, 2011a, 2011b)	42
Figure 3.	Ecological Métissage	72

PRELUDE: A TALE OF TWO FIRST NATIONS

Two Aboriginal[1] communities in western Canada, Saddle Lake First Nation in Alberta and T'Sou-ke First Nation on Vancouver Island, British Columbia, have recently been capturing the attention of both Aboriginal and non-Aboriginal leaders, scholars, and educators due to innovative environmental initiatives and associated social, economic and educational developments.

In Saddle Lake a revolutionary, low-impact water treatment system has attracted interest from other remote Aboriginal communities (Narine, 2009). Saddle Lake was once a highly polluted body of water; community members had to boil all water prior to consumption due to inadequate waste and fresh water treatment facilities (Drake, 2006). After repeated appeals, the community received financial assistance from the federal government to clean up the lake and develop a new water treatment system. Against the advice of government and industry experts, Saddle Lake partnered with researchers from the University of Alberta to develop a revolutionary water treatment system. The project managers were influenced by the vision of their Elders to embrace modern science guided by traditional wisdom. As one participant commented, "It's always been a desire of the Elders to embrace sound science and…traditional holistic teachings to fashion healthy drinking water" (Narine, ¶ 6).

Rather than using chemicals to treat their water, further tampering with the highly disturbed ecosystem, the Saddle Lake team developed a system that uses a non-invasive integrated membrane filter (Narine, 2009). Saddle Lake's treatment system has been highly successful and is now in high demand across Canada. The project's managers are presently busy sharing the technology with other Aboriginal communities.

T'Sou-ke First Nation, in cooperation with industrial and governmental partners, also made headlines recently for developing the largest solar energy project in British Columbia in addition to other sociocultural initiatives (CBC, 2009; Kimmett, 2009; Ozog, 2012). T'souke's solar committee members were motivated by a desire to become a self-sufficient community capable of providing their own energy and food; they successfully collaborated with renewable energy experts to make their vision a reality (Kimmett, 2009).

Located on the southern shores of Vancouver Island, the community is now able to generate enough energy to power 25 homes and several administrative buildings. At times they generate so much electricity that they are able to profit by redirecting surplus power back into the provincial grid. The community has also installed solar hot water tanks on some homes and buildings and plans to develop wind power and an organic farm in the future (CBC, 2009).

Nine Band members also underwent training with First Power, a solar power developer that employs Aboriginal pedagogical approaches, to become certified solar panel installers. As one participant described, "We learned by doing. They took a group to each house and would teach us on an actual system. Having everything right there in front of you made it a lot easier" (Kimmett, 2009, ¶ 20–21).

Chief Gordon Planes commented that the project was guided by the traditional Potlach mentality of generosity; they are already actively sharing their experiences with other communities, Aboriginal and non-Aboriginal alike, and have become an inspiring example for others to follow (CBC, 2009; Ozog, 2012). As Kimmett (2009, ¶ 5) notes, "supporters call it the beginning of a renaissance for First Nations; a way to connect with the land in a totally new way plus gain energy autonomy." Henderson (2013) also enthusiastically suggests that Indigenous communities across Canada are increasingly becoming leaders in renewable energy and other ecologically conscious initatives.

Might such examples provide a vision for the future of environmental education in Canada? In such a context, how might we authentically and respectfully combine Indigenous, Western and other culturally rooted traditions of science and technology to address contemporary socio-ecological and educational issues for the benefit of all Canadians, Indigenous and non-Indigenous alike?

Note

1. In order to recognize and honour the importance of certain terms and concepts, I follow Métis scholar Fyre Jean Graveline's (1998) example by capitalizing them in this book. These terms include: Indigenous, Aboriginal, First Nations, White, European, Western, Métis, Land, Earth, and Nature.

. 1 .

INTRODUCTION

Motivated by recent environmental initiatives such as Saddle Lake's water treatment system and T'Sou-ke First Nation's solar energy project and inspired by examples of cultural métissage in Canadian history, the purpose of the doctoral study that led to the development of this book was to explore "ecological métissage" as an emerging vision for environmental education in Canada (Lowan-Trudeau, 2012b, 2013b). The concept of ecological métissage arises from Thomashow's (1996) description of "ecological identity," as the way that we understand ourselves in relation to the natural world, and an understanding of "métissage" as the mixing or blending often associated with culture or ethnicity (Chambers, Donald and Hasebe-Ludt, 2002; Nguyen, 2005; Pieterse, 2001). For the purposes of this inquiry, ecological métissage denotes a blending of two or more ecological worldviews at a personal and/or cultural level as represented in personal identity, philosophies, and practices.

This theme is also supported by an increasing number of scholars and educators who advocate for the integration of Indigenous, Western and other knowledges in our collective attempts to address the world's current ecological crises. Indigenous and non-Indigenous environmental educators alike are working to bridge cultural gaps as well as to revive and preserve Indigenous traditions and ecological knowledge, ever conscious of the delicate balance between respectful sharing and misappropriating or misusing Indigenous knowledge.

However, as our field has grown, divergent perspectives have emerged regarding the current and potential relationship between Indigenous, Western, and other culturally based ecological knowledge systems and philosophies. While some believe that Western and Indigenous knowledge systems should never be blended (Kimmerer, 2013), carefully maintaining mutually respectful separate existences that may benefit and interact with each other from time to time, others such as the Integrative Science Institute (2012), a collaboration between Indigenous and non-Indigenous researchers at Cape Breton University, Canada advocate for "Two-Eyed Seeing", a more comprehensive integration of Western and Indigenous science (Bartlett, 2005; Hatcher, Bartlett, Marshall and Marshall, 2009; Integrative Science Institute, 2012). The Integrative Science Institute also proposes the possibility for "Three-" or "Multiple-Eyed Seeing", simultaneously viewing and addressing contemporary ecological issues from Western, Indigenous, and other cultural perspectives. Projects such as this and encouragement from participants in my doctoral study led to a follow-up pilot study into the pedagogical experiences of newcomers to Canada regarding Indigenous ecological knowledge in Western science contexts. Insights from both studies are presented in this book.

I understand the reservations of those concerned about the potential for Indigenous knowledge to be misunderstood and subsumed if integration is not approached with extreme care. However, as a Métis Canadian of mixed Algonquian and European ancestry on both sides of my family, my personal experiences and reflections upon past and present happenings in Canada and beyond led me to the optimistic position at the outset of this inquiry that bringing Western, Indigenous, and other ecological knowledges and philosophies together in a highly integrated manner is possible and, in fact, vital to the future survival and well-being of our communities. However, in the course of these two studies I developed a more cautious attitude. Such tensions and considerations formed the impetus for these two studies and, as will be presented throughout this book, I had the wonderful opportunity to discuss and critically consider these and other related issues and concepts with experienced Indigenous and non-Indigenous environmental educators, community leaders, and newcomers across Canada. However, in keeping with Indigenous protocols, prior to discussing these two related studies further I will tell you about myself.

Personal Background: Positioning Myself

As Absolon and Willet (2005) explain, positioning, or introducing oneself in detail, is an important aspect of relationship building in Aboriginal contexts. Similar to the Cherokee-American writer Owens (2001, p. 11), I am "a person of deeply mixed heritage and somewhat unique upbringing." I am a proud Métis with Indigenous and European ancestry from both my mother (Mi'kmaq, French, Celtic, and Norwegian) and father (Lenape, Swiss, and Celtic). Similar to many Métis families, these ancestral streams mixed, mingled, and meandered along the East coast, through the central woodlands, and over the prairies of North America over many generations.

Growing up on the foothills and eastern slopes of the Rocky Mountains in the city of Calgary, where the Bow River meets the Elbow River, I was raised conscious of my family history. Tales of Métis ancestors using traditional plant knowledge to cure neighbours' ailments intermingled with stories of Norwegian relatives singing Lutheran hymns of gratitude as rain nourished their parched Saskatchewan grain fields. This métissage of family stories has always intrigued me.

Despite being raised in a city, my childhood was a constant métissage of urban and rural environments; I was fortunate to spend a considerable amount of time exploring the mountains, forests, rivers, and coastline of western Canada as a child. My parents took every opportunity to leave the city for our cabin in central Alberta, the nearby mountains, or to visit relatives on British Columbia's west coast. During my elementary school years, I was often removed from school to accompany my father on outdoor education trips that he led for junior high school students. As I grew older, I participated in countless sailing, hiking, fishing, canoeing, orienteering, and skiing excursions both with my family and school. Looking back, I realize how those early experiences deeply influenced my personal and professional journey.

Over the past fifteen years, my professional experiences have been based primarily in educational environments. After completing undergraduate studies, I spent a year teaching high school English in rural Japan before returning to Canada to begin my career as an outdoor and environmental educator in earnest. The past ten years have seen me facilitating an outdoor pursuits program for adults with developmental disabilities, working with Aboriginal youth on extended expeditions in the northern boreal forest, teaching and guiding youth and adults in the foothills and eastern slopes of the Rocky

Mountains by ski and canoe, and teaching in several post-secondary institutions across Canada.

My family, academic, and professional experiences have all influenced my growth and development as an educator. I drew on them in crafting this study while attempting to maintain a reflexive self-awareness of my own role in and influence on the research journey.

In this book I take the position of a Métis Canadian exploring the relationship between Indigenous and Western understandings of, and interactions with, Nature. As a Métis person, I find the concept of ecological métissage especially intriguing, however, my intention is not to be exclusive—my experience is that there are a growing number of people from many cultural backgrounds including First Nations, Métis, and non-Aboriginal who embrace a similar perspective in Canada and around the world.

Considering the Canadian Context

This journey was grounded in the Canadian context and my perspective as a Métis scholar and educator. However, with due consideration for the regional nuances of Indigenous and ecological issues, I believe that readers from other nations will also find significant resonance and insight.

Canada: A Métis Nation?

During the early stages of this study I was intrigued by the work of Canadian philosopher Saul (2008). In *A Fair Country*, Saul challenges the notion of Canada being founded by only two nations (French and British) and proposes that we are, in fact, a "métis" nation. Saul emphasizes the long forgotten and largely ignored Indigenous foundations of Canadian society. He suggests that Canada is, in fact, a culturally and linguistically "métis" society. He states:

> We are a métis civilization. What we are today has been inspired as much by four centuries of life with the indigenous civilizations as by four centuries of immigration. Perhaps more. Today we are the outcome of that experience. As have Métis people, Canadians in general have been heavily influenced and shaped by First Nations. We still are. We increasingly are. This influence, this shaping is deep within us. (p. 3)

Saul suggests that for the sake of our collective Canadian ecological, political, linguistic and socio-cultural identities, we must dust off centuries of denial and embrace our common Aboriginal heritage. He proposes that a cursory examination of our justice system, political structure and history, as well as Canadian French, English, and Aboriginal languages will quickly reveal the inextricable influence of Canada's founding cultural groups on each other. Saul suggests that living with cultural complexity was, and continues to be, one of Canada's unique strengths. We inherit this from our Aboriginal forebears who had learned to navigate a complex multicultural landscape long before the arrival of the first Europeans. In his introductory comments, Saul proposes that:

> A dancer who describes himself [sic] as a singer will do neither well. To insist on describing ourselves as something we are not is to embrace existential illiteracy. We are not a civilization of [purely] British or French or European inspiration. We never have been...To accept and even believe such fundamental misrepresentations of Canada and Canadians is to sever our mythologies from our reality...We are a people of Aboriginal inspiration organized around a concept of peace, fairness and good government. That is what lies at the heart of our story, at the heart of Canadian mythology, whether francophone or anglophone. If we can embrace a language that expresses that story, we will feel a great release. We will discover a remarkable power to act and to do so in such a way that we will feel we are true to ourselves. (pp. xi–xii)

As a Canadian of mixed Aboriginal and European heritage, I find great resonance with Saul's thesis. However, Saul's comments and his liberal use of the term "métis" might seem overly optimistic to some due to the undeniably negative impact of European colonialism on Indigenous cultures, languages, and health. The effects of colonization were and continue to be devastating and wide-ranging for Aboriginal peoples (Adams, 1999; Bastien, 2003; Battiste, 1998, 2005; Simpson, 2002; Graveline, 1998). Colonization has disrupted cultures, destroyed languages, and dislocated Indigenous peoples around the world from their languages, traditional lands, and ancient practices. However, as Battiste (1998) suggests, Indigenous peoples are resilient and we are now witnessing the re-emergence and rebuilding of Indigenous communities. As part of this healing process, many Indigenous leaders are facilitating the revival of traditional knowledge not only for the benefit of Aboriginal peoples, but also for the rest of society (e.g., Barnhardt and Kawagley, 2005; Snow, 1977/ 2005). Barnhardt and Kawagley (2005) state:

> Indigenous peoples throughout the world have sustained their unique worldview and associated knowledge systems for millennia, even while undergoing major social upheavals as a result of transformative forces beyond their control...The depth of

> Indigenous knowledge rooted in the long inhabitation of a particular place offers lessons that can benefit everyone, from educator to scientist, as we search for a more satisfying and sustainable way to live on this planet. (p. 9)

As Barnhardt and Kawagley (2005) suggest, Indigenous peoples developed highly sophisticated and intimate understandings over thousands of years of this land that we all presently inhabit. Amidst the plethora of contemporary ecological concerns, wouldn't it make sense to deeply consider Indigenous knowledge and perspectives? Saul's (2008) thesis may be seen as an invitation for us to revisit history, acknowledging the wrongs that were committed, but also recognizing inspiring examples of intercultural co-operation and métissage, moving forward to collectively re-imagine our cultural and ecological identities.

Canadian Ecological Identity

One might justifiably question if there is such thing as a "Canadian Ecological Identity". Our national symbols and icons—the beaver, the canoe, the voyageur, the lumberjack, the maple leaf, the Group of Seven, the Great Lakes, the Rocky Mountains, the North—suggest that, as a nation, we identify strongly with the natural world (Francis, 2005). Foreign tourists flock to Canada annually to experience our abundance of natural splendour. Why is it, then, that we recently received a lifetime unachievement "Colossal Fossil" award for being the least environmentally progressive nation in the world (see www.climateactionnetwork.ca); might there be a contradictory element to our national consciousness?

Adrift in our National Consciousness: Meditations on the Canoe

In his reflections on Canadian identity, satirist Ferguson (2007) suggests that:

> Canada is a land pinned between the memories of habitant and voyageur. We have grown crops and built cities, bypassed rapids, unrolled asphalt and smothered our fears under comforters and quilts. We are habitants, and the spirit of the voyageur now lingers only in the home movies of our nation…Like a song from the far woods. (p. 94)

Might Ferguson, though somewhat glib, be correct? Has mainstream Canadian society happily adapted to modernity along with the rest of the Western world, while desperately grasping for increasingly distant images of Nature as touchstones for an increasingly urban existence? Might we be much more disconnected from the natural world as a society than we would like to admit?

I must confess to romantic visions of my own of adventurous voyageur ancestors setting out each spring from the comfort of their *habitant* farms on the southern shores of the St. Lawrence River to spend the spring, summer, and fall plying the waters of the Great Lakes and the rivers of the Northwest. Perhaps these dreams manifested themselves in my own passion for recreational canoe travel? Is that all that I am doing when I set out on a three-day, one-week, or even one-month canoe trip? Recreating? Re-creating? Or is there a deeper meaning to these intermittent adventures?

Like Cole (2002, p. 450), I believe that "my canoe is a place of cultural understanding"; when paddling, I often reflect on the Indigenous roots of the canoe. I think back to my Mi'kmaq and Lenapé ancestors, skilled canoe builders who deftly created sea- and river-worthy boats from birch, cedar, spruce, and other trees in the eastern woodlands; this gives me a profound sense of connection to the Land and my own cultural history. There are also moments when I experience the "flow" that Czikszentmihalyi (1990) describes—a feeling of oneness with the Land when you feel yourself lost in the moment and time slows down. And sometimes I simply enjoy paddling as a physical and/or social activity that I can do with my friends, family, and students.

However, Dean (2006; 2013) critiques the role of the canoe as a celebrated and arguably misappropriated icon of mainstream Canadian culture. For example, during the Centennial celebrations of 1967 the Canadian government organized the longest canoe race ever held. Teams of paddlers from every province and territory retraced the historic route of the voyageurs from Rocky Mountain House, Alberta, to Montréal, Québec (Dean, 2006; Guilloux, 2007). However, Dean suggests that the Centennial canoe race was a misrepresentation of Canadian history. She notes that while most voyageurs during the fur trade were French Canadian, Aboriginal, or later Métis, the large majority of Centennial paddlers were English-speaking Euro-Canadians. Dean also relates that the Aboriginal participants in the Centennial race were often poorly treated. She suggests that the Centennial canoe race was an instance of dominant Anglo-Canadian society appropriating a cultural icon that is not really their own. Why would they do this? What is it about the canoe and other symbols of our nation's connection to Nature that so strongly captures the imagination of the average Canadian? Are we clinging to the past, whether real or imagined, or seeking solace from the present?

The Promise of Escape

Novelist MacGregor (2003) suggests that a large characteristic of Canadian identity is based upon the notion of "escape". He proposes that throughout history up to the present day, Canada has been seen as a place of escape for refugees or immigrants fleeing poverty or violence in their homelands as well as a haven for romantic wanderers or idealists seeking isolation. MacGregor also presents the notion that escape is preserved in "cottage country" throughout Canada and the annual pilgrimages that so many Canadians make to their favourite fishing, hunting, canoeing, or skiing destinations to "get away from it all". The escape mentality uncritically presented by MacGregor (2002) portrays Nature as an isolated refuge from the "real world", similar to the interpretation of the Western concept of "wilderness"[1] as a place of solace or retreat (Merchant, 2004).

I believe that this kind of attitude represents an ecological identity that, while reverent, views Nature as a recreational *resource*, useful for a short period of time to recharge before returning to the rigours of city life. While the escape mentality may not be immediately harmful on the surface, I believe that it is symptomatic of a disconnected Nature-as-resource mentality that is ironically used often by urban preservationists to critique the actions and attitudes of rural conservationists or resource extractors (Berry, 2009; Thomashow, 1996). While the immediate effects of resource extraction are much more obvious, is there really that much difference in the original mentality? In both cases, the more-than-human world is ultimately viewed as a commodity available for human use and manipulation, as long as it suits us. Having lived in both large metropolitan centres as well as isolated rural and semi-rural areas, it is my experience that rural farmers, hunters, loggers, miners, and fisher folk are often much more keenly aware of and deeply connected to the Land around them than the urban environmentalists who so often criticize and dismiss them with scorn. As American farmer and eco-philosopher Berry (2009) astutely observes:

> They have trouble seeing that the bad farming and forestry practices that they oppose...are done on their behalf, and with their consent implied in the economic proxies they have given as consumers. (p. 78)

When considering ecological identity in Canada, the picture is often unclear. For example, contrary to popular perception of the province of Alberta as the home of unabated oil and resource extraction, a recent survey into the environmental attitudes of Albertans reported that a majority of people in the province actually hold positive feelings towards the "environment", but most feel disempowered or at a loss to act or speak out (Thompson, 2009). Statistics Canada (2008) also reported that "the environment" was the top concern for Canadians in 2007. My hope is that these studies are examples of a slow shift in our society that is increasingly positively disposed towards environmental issues.

However, despite our cherished national image as a naturally beautiful and environmentally pristine nation, Canada's current government has been awarded the infamous "Colossal Fossil" award five years in a row and a lifetime "unachievement" award by the Climate Action Network (2013) for being the least environmentally progressive nation in the world. What happened to all those Canadians who ranked "the environment" as their top concern in 2007 (Statistics Canada, 2008)? Have their priorities shifted due to the recent worldwide economic downturn? Or, perhaps, our federal government simply does not represent the interests and values of a majority of Canadians?. The most recent federal election where the Conservative Party was elected to a majority government with only forty percent of the popular vote (Elections Canada, 2011) would suggest that the latter may indeed be the case.

What do all of these statistics really mean? Perhaps a political and/or economic crisis is exactly what is needed for the growth of ecological métissage—an opportunity to reassess and re-imagine our society. Perhaps, like post-modern voyageurs, we have ventured deep into the wilderness of industrialization and modernity, only to realize that we don't have the tools, skills, and wisdom to survive. Might society at large turn to the wisdom of Indigenous peoples to reassess how to live well in our local places? How would this be accomplished in theory and/or practice?

La Metis des Grecs: A Trickster Tale

In the early stages of my doctoral studies, I was also intrigued by the Ancient Greek concept of *metis*[2] (pronounced "meh-tiss"), a subtle, oblique, and intuitive form of knowledge that was once widely recognized and celebrated but eventually suppressed and ignored by Western societies due to its associations with femininity and Nature (Dolmage, 2009). Détienne and Vernant (1974, 1991) suggest that:

> There has been a prolonged silence on the subject of the intelligence of cunning [metis]...from a Christian point of view, it was inevitable that the gulf separating men [sic] from animals should be increasingly emphasized and that human reason should appear even more clearly separated from animal behaviour than it was for the ancient Greeks. (pp. 318–319)

Similar to the Trickster figures in many North American Aboriginal cultures (Graveline, 1998; King, 2003; McDermot, 1993; Reid and Bringhurst, 1996), examples of metis in Greek mythology and philosophy often involve the *dolos* (tricks or ruses) of animals such as the fox, the octopus, or the squid, who is able to turn itself inside out.

As a Métis person, I was struck by the etymological similarity of Métis with the Greek term *metis*. Dolmage (2009) states:

> The French word *métis* is related to the Spanish word *mestizo*, both coming from the Latin word *mixtus*, the past participle of the verb *to mix* and connoting mixed blood. In critical theory the concept of *métissage* also locates and interrogates the ways that certain forms of knowledge have been relegated to the margins... *métissage*...etymologically linked to *mêtis* and meaning mixture or miscegenation, has been used as a critical lens through which one might observe issues of identity, resistance, exclusion, and intersectionality. (pp. 24–25)

Dolmage also suggests that:

> The form of the word itself [metis] is a kind of trick: the Greek words *me* and *tis* mean "no man" or "no one". But the two words put together label a particular someone: the sort of person whose identity can be elusive, who is unpredictable but resourceful and clever. (p. 5)

Dolmage's etymological observations resonate strongly with my own thoughts and experiences as a Métis Canadian. Being a bi- or multicultural person has distinct advantages, but it may also create the feeling that Dolmage alludes to of being "no one", someone who doesn't quite fit in a particular social or cultural milieu, but somehow seems to function effectively anywhere.

Throughout Canadian history, Métis peoples have often been intercultural mediators, deftly navigating between and mediating European and Aboriginal cultures (Saul, 2008). Many Métis embodied the Greek concept of metis in their intuitive understanding of the dynamic relations between the myriad of European, Aboriginal and other cultures that came together to form what we know today as Canada.

Explorations into the Greek concept of metis and reflections upon Saul's (2008) proposal that Canada is a historically "métis nation" led me to wonder what characterizes intercultural environmental educators in Canada today? Who are these cultural "border crossers" (Hones, 1999; Nguyen, 2005; Pieterse, 2001)? What led them to their chosen vocation? What makes them effective? And how might they be reshaping Canadian ecological identity? I was also guided by the overarching question, "*Can Western and Indigenous knowledge of the natural world be blended theoretically and in practice? If so, how?*" and the additional guiding questions:

1. What characterizes the ecological identities of contemporary intercultural environmental educators?
2. Do they embody ecological métissage? If so, how?
3. How might the concept of ecological métissage reshape environmental education in Canada?

As presented and discussed in the following chapters, these questions provided an initial framework to explore and expand upon controversial and under-explored issues and concepts in contemporary intercultural environmental education in Canada. They also provided me with a guide to interpret the experiences and perspectives of the intercultural environmental educators who participated in the first, my doctoral, study.

Motivated by insights from participants in the first study related to the rich complexity encountered by educators and learners in multicultural settings, the purpose of the second short follow-up study was to explore the formal and informal educational experiences of first generation immigrants to Canada with Indigenous ecological knowledge and philosophy in Western science contexts. Specifically, I was guided by the following questions:

- How do newcomers to Canada perceive Indigenous ecological knowledge and philosophy?
- How might formal and informal science and environmental educators better respond to such culturally complex educational contexts?
- What are the broader societal implications of these kinds of questions?

These research journeys produced profound dialogues and meta-dialogues between and with the participants and the literature. However, it is centrally important to note that these are only *my* interpretations; it is appropriate and anticipated that other readers (listeners) might have different interpretations.

Notes

1. "Wilderness" is a contested term that carries strong Western, anthropocentric connotations (Merchant, 2004). It implies a mythically forbidding untamed "otherness", threatening to humans, but alluring nonetheless (Merchant, 2004; Snyder, 2003). Many Indigenous scholars also challenge the use of "wilderness" to describe areas relatively undisturbed by humans because, for Indigenous peoples, "the Land", which includes all the physical and metaphysical elements of Creation, is not viewed as wild and forbidding, it is, as Snyder (2003) suggests, home (Cajete, 1994; Snow, 1977/2005). In congruence with this perspective, I avoid the use of the term "wilderness" in favour of terms such as "Nature" or "the Land" to describe the natural world in this book.
2. In this study "Métis" refers to the Métis people of North America, while "Metis" will be understood as a figure from Greek mythology, with "metis" denoting a recognized form of knowledge in ancient Greek society.

. 2 .

METHODOLOGICAL MÉTISSAGE

As previously described, this book weaves together findings from two related studies. In the course of the first study, which led to my doctoral dissertation, I had the privilege to interview ten Indigenous and non-Indigenous intercultural environmental educators from a variety of backgrounds (Sto:lo, Métis, Pakistani, Japanese, various European cultures) from across Canada who draw upon Western, Indigenous, and other cultural traditions to inform their ecological identities, philosophies, and practices. Motivated by insights from participants in the first study, the second comprised a pilot project that explored the experiences of newcomers to Canada with Indigenous ecological knowledges and philosophies in Western science pedagogical contexts.

I was guided through both of these research journeys by a methodology that I developed from an initial bricolage (Berry, 2006; Steinberg, 2006) or integration of Indigenous (Kovach, 2010; Smith, 1999; Wilson, S., 2008) and interpretive (Berry, 2006; Denzin, 1989; Steinberg, 2006; Willinsky, 2006) methodologies to form a new methodological métissage (mix) representative of my own identity as a Métis scholar and educator.

From Bricolage to Métissage: A Process

Berry (2006) introduces the concept of bricolage as methodology by sharing an anecdote about an Acadian friend who is constantly at work using "scraps of leftover wood…to create the most unique and charming birdhouses…no two ever look the same" (p. 87). Berry notes that, like her friend's carpentry projects, engaging with bricolage as a research approach involves working "with 'bits and pieces' of theoretical, methodological and interpretive paradigms. It works with the scattered parts, overlaps and conflicts between paradigms" (p. 102). Steinberg (2006) also comments that:

> Bricolage involves taking research strategies from a variety of scholarly disciplines and traditions as they are needed in the unfolding context of the research situation. Such an action is pragmatic and strategic, demanding self-consciousness and awareness of context from the researcher. The *bricoleur*, the researcher who employs bricolage, must be able to orchestrate a plethora of diverse tasks. (p. 119)

I have come to realize that I was acting as a bricoleur at certain stages of this inquiry as I drew upon a diversity of cultural and academic sources such as Western and Indigenous theorists, historians, scientists, educators, and interpretive qualitative researchers from around the world. As Steinberg (2006) notes, and as I experienced first-hand, the methodological diversity inherent in bricolage-inspired studies is not selected at random due to a lack of organization or focus. This methodology is a carefully considered, dynamic, and intuitive attempt to express my own identity as a Métis researcher and engage with a diversity of voices in order to foster rich and respectful conversations between myself, the literature, and the research participants.

While the concept of bricolage denotes the calculated cobbling together of various elements, resulting in a dynamic, but ultimately deconstructable whole (Berry, 2006; Roth, 2008; Steinberg, 2006), Roth suggests that, in cultural terms, bricolage often leads to métissage, a term implying a mix or blend so complete that the parts can no longer be extracted from the whole.

Métissage as Methodology

I was also influenced in the early stages of developing this methodology by scholars such as Chambers, Donald, and Hasebe-Ludt (2002) who characterize métissage as "a...way of merging and blurring genres, texts and identities...a creative strategy for the braiding of gender, race, language and place into autobiographical texts" (pp. 1–2). Donald (2009) describes a researcher employing métissage as the weaver of a braid or a Métis sash expressing, "the convergence of wide and diverse influences in an ethically relational manner" (p. 142).

Another metaphoric image that speaks to this intertwined Western-Indigenous relationship is the infinity symbol found at the centre of the Métis flag. As Dorion and Préfontaine (1999) note, "the horizontal eight is an infinity sign, which has two meanings; the joining of two cultures [European and Indigenous] and the existence of a people forever" (p. 17).

My adaptation of métissage as methodology employs the infinity symbol as a metaphoric image (see Figure 1 below). It is similar in inspiration to Donald's (2009, 2010) because it critically compares and combines both Western and Indigenous traditions. However, while Donald's application of métissage involves comparing and contrasting colonial and Indigenous narratives of historical sites and objects, my approach focuses on a métissage of methodological influences that explore contemporary peoples' lives, experiences, and perspectives through a narrative approach.

Research in the Third Space: Comparing Indigenous and Interpretive Inquiry

As I began my work as a methodological bricoleur, I set out to clarify the relationship between Western and Indigenous approaches to research as they related to my project. I discovered that, when comparing traditional Indigenous approaches to gathering and transmitting knowledge (Barnhardt and Kawagley, 2005; Kovach, 2010; Smith, 1999) to Western science-inspired research models such as quantitative and positivistic approaches (Creswell and Miller, 2000), some similarities may exist such as "empirical observation in natural settings, pattern recognition, verification through repetition, [and] inference and prediction" (Barnhardt & Kawagley, p. 16), but overall there is significantly more *divergence*. For example, Barnhardt and Kawagley emphasize that while Western science-rationalist epistemology takes a reductionist view of the universe, attempting to isolate and analyze specific parts, Indigenous cultures recognize a holistic interconnection of all Creation (humans, more-than-humans, living and non-living entities in nature). Barnhardt and

cultural traditions through methodology, but for all researchers, due to the fundamental fact, as previously stated, that all environmental education work and research in Canada is conducted on the traditional territory of Indigenous peoples (Wilson, S., 2008). In the following, I explore each of the concepts presented in the table and figure above as they relate to my study and the field of environmental education research.

Narrative Inquiry and the Oral Tradition

> Our theory of knowledge is found in the sacred stories that are the living knowledge of the people. The stories explain the nature of reality, the science, and the economic and social organization of *Siksikaitsitapi*. They are the accumulated knowledge of centuries. Each generation…is responsible for retelling the stories to the next generation. The knowledge contained in them is living. (Bastien, 2003, p. 45)

As Blackfoot scholar Bastien relates above, the oral tradition is a foundational characteristic of Indigenous cultures. Due to the centrality of the oral tradition and story-telling in Indigenous cultures, Indigenous scholars such as S. Wilson (2008) and Kovach (2010) suggest that narrative methodology, as an interpretive approach, is a relevant and appropriate methodology for Indigenous research, even though it involves presenting stories in writing.

The oral tradition has existed in many societies for thousands of years as a way to preserve history, family lineages, and cultural stories and values (Finnegan, 1970/1996; Hart, 2002). Snow (1977/2005) states that the oral tradition is still taken very seriously in Indigenous communities: for example, contracts and agreements negotiated orally are accepted as lawful and binding by Indigenous peoples. Kovach (2010) and Miller (2011) also note that oral traditions are increasingly respected in the Canadian judiciary as evidence in land-claim cases. Miller comments that, "In the 1980s and 1990s courts and tribunals in Canada and other jurisdictions began to seriously consider the relevance of Aboriginal oral narratives…to legal proceedings" (p. 2). Kovach also relates the well-known, "Delgamuukw decision, [where] the Supreme Court of Canada ruled that oral testimony has the same weight as written evidence in land entitlement cases" (p. 95).

Interpretive scholars such as Vansina (1961/1996) and Indigenous scholars such as S. Wilson (2008) both suggest that the oral tradition is dynamic and must be considered in the context from which it arises. They note that the oral tradition is an interactive record of perceptions that links lives to context. One of its strengths is that it is highly adaptable; for example, a story meant to pass on cultural values might be updated and interpreted to suit the current lives of its audience without changing the original meaning or lesson of the story. In this manner, the audience might better relate to the story as it comes alive in their contemporary world.

Narrative Inquiry

Hart (2002) notes that narrative inquiry, which grew out of the oral tradition, is increasingly employed in environmental education research and qualitative inquiry in general. He suggests that it is "as much a way of knowing ourselves as a way of organizing and communicating the experiences of others" (p. 143).

Kovach (2010) and Denzin (1989) both comment that narratives may cover broad life histories or more specific topics. For example, my doctoral study was not simply an open biographical exploration of the participants' lives: I focused on the development of their cultural and ecological identities and their experiences with, and beliefs about, topics such as the relationship between Indigenous ecological knowledge and Western science. In order to create a trustworthy "portrait" (Lawrence-Lightfoot, 2005) and contextualize the perspective of each participant, I provided background information and significant amounts of dialogue on the interview setting and flow. I did not construct completely comprehensive chronological accounts of the participants' lives, rather the focus of each mini-biography was the subject matter or issues as identified in my original research questions with special attention to "epiphanic" or "aha" moments (Denzin, 1989).

Clandinin and Connelly (2000), well-known narrative methodologists, remind us of the embeddedness (socially, culturally, historically) and continuity of experiences in our lives and that significant experiences and epiphanic moments through stories can provide us with enhanced insight into other peoples' lives. It is also important to consider that life is a continuous series of interrelated events situated in various contexts, the impact of sustained, but perhaps less exciting, experiences and the contexts within which they occur.

In keeping with the oral tradition (Wilson, S., 2008), I framed the final discussion in both of these studies by referring back to my original research questions, weaving intriguing ideas and responses shared by the participants into a dialogue between them, the literature, and myself. I also highlighted epiphanic moments that I experienced or witnessed with the participants.

Reflexivity

Many scholars in the areas of interpretive methodologies (Steinberg, 2006; Tobin, 2006), Indigenous research (Kovach, 2010), and environmental education (Lotz-Sisitka, 2002) stress the critical importance of reflexivity. A reflexive researcher examines their role in the research process, reflecting on their experiences throughout the research journey, the influence of their cultural and social positioning, and their interpersonal interactions with research participants. Explicit reflexivity is also a tool for demonstrating the learning experienced by the researcher, a key criteria for quality interpretive research (Tobin, 2006). Reflexivity recognizes that a qualitative researcher is also a participant in the research process. As Kovach (2010) explains from an Indigenous perspective:

> In co-creating knowledge, story is not only a means for hearing another's narrative, it also invites reflexivity into research. Through reflexive story there is opportunity to express the researcher's inward knowing. Sharing one's own story is an aspect of co-constructing knowledge from an Indigenous perspective. (p. 100)

Steinberg (2006) and Tobin (2006) also suggest that reflexivity is a vital element of interpretive research because it demonstrates the learning experienced by the researcher. I employed reflexivity and demonstrated learning in both of these studies as I narrated and responded to the literature, and my changing perceptions and understandings of key concepts through conversations with participants.

Berry (2006) also notes that reflexivity is very important in interpretive research because it allows and encourages the researcher and the participants to position themselves theoretically, culturally, geographically, and ecologically, another key aspect of both interpretive environmental (Berry, 2006) and Indigenous research (Absolon & Willet, 2005).

Positioning

Absolon and Willet (2005) suggest that explicitly positioning yourself is an especially important aspect of reflexivity in research projects involving Aboriginal peoples because positioning is an integral foundation of many Indigenous cultures. From an interpretive research perspective, positioning means introducing yourself to your research participants and later your audience (Bolak, 1997). For example, at the outset of any interview or conversation,

I explicitly inform my participants that I am Métis on both sides of my family with deep roots in the prairies and northeastern woodlands of Turtle Island (North America). This helps them to understand my perspective and background and allows them to position themselves in response. Participants in several studies, Aboriginal and non-Aboriginal alike, have commented that they felt more at ease discussing culturally related topics knowing my background because they had a better sense of their audience (i.e., me).

Conscious of such nuances, Ginsburg (1997) also describes the importance of situating not only the interviewer but also the interviewee; relating details about the context of the interview, such as the physical setting, also adds richness to the final product. I also did this in both of these studies for each participant by including a significant amount of biographical information as well as by initially presenting participants' interviews as intact narratives rather than breaking and mixing them into themes. This is an appropriate protocol in Indigenous research where treating participants and participants' stories with respect is of the utmost importance (Bastien, 2003; Kovach, 2010). Reflexive researchers understand research to be collaborative, interactive, and constructive. Researchers and participants work together, consciously and unconsciously, to create and interpret the interview/research experiences. Positioning yourself and your participants theoretically, geographically, and ecologically is also an important component of reflexivity, especially in Indigenous and environmental research.

Problematizing: Positioning Yourself Theoretically

Grele (1994) and Steinberg (2006) remind us that an interview is a conversational narrative, not only between the interviewer and the participant, but also with the literature of their field that is embedded in current and historical contexts. Grele also recognizes that reflexive research can lead to conflict and tension with participants. This is the natural product of people who might hold conflicting worldviews or experiences interacting. Interviews often touch on controversial or sensitive topics and this can lead to conflict if differing perspectives exist between the interviewer and interviewee. Recognizing, reflecting upon, and reporting on this tension can deepen the research process and lead to richer results.

I embraced Grele's suggestions and responded to participants in these studies with my own thoughts on key issues as well as those of other scholars and participants in a cautious manner. This typically resulted in a deepening of our discussion; one participant even commented that our conversation was going to "keep her up all night" thinking further about the relationship between her identity and practice as a Métis environmental educator working with Aboriginal and non-Aboriginal youth.

In consideration of interpretive scholars such as Berry (2006), who explains that problematizing key historical and sociocultural issues relevant to a respective study is a key preparatory step for interpretive research, I problematized historical and sociocultural concepts and topics throughout both of these studies. For example, in the following chapters, I review and discuss challenging and controversial concepts such as juxtaposing various Western philosophical traditions (e.g., Western science, deep ecology, and bioregionalism) with Indigenous approaches as understood from my own perspective as a Métis Canadian. I also challenge popular historical and contemporary notions of Métis peoples and cultures.

Contextualizing: Positioning Yourself Geographically and Ecologically

Bastien (2003, p. 42) notes that responsible citizenship in Indigenous cultures means, "taking care of each other *and* our environment." This sense of relationship and reciprocity with the greater than human world is foundational to Indigenous cultures and research (Bastien, 2003; Cajete, 1994) and it bears promising similarity to the beliefs of many contemporary Western environmental philosophers, educators, and researchers (Naess and Rothenberg, 1990). Cajete also emphasizes the localized specificity of Indigenous cultures, emphasizing the importance of recognizing the unique relationship that every culture develops to a specific geographical area over thousands of years.

As a result, positioning yourself geographically and ecologically is a key aspect of both interpretive and Indigenous research that provides a clear link to environmental education contexts. I satisfied this important condition by situating the research participants, the literature, and myself socio-culturally, geographically, and historically throughout both studies. All of the research participants were also situated geographically and ecologically in their biographies through introductory comments and the sharing of their personal philosophies and experiences.

I also attempted to embody both deep and shallow ecological perspectives (Naess & Rothenberg, 1990) in my research methods. I found this to be very satisfying, but it also presented tensions at times. For example, in order to limit my use of fossil fuels, I conducted some interviews using video conferencing software. This presented an interesting dilemma because at the beginning of interviews conducted in-person, I always offered tobacco to participants, a common practice amongst most Indigenous peoples across North America for requesting the sharing of knowledge (Kovach, 2010; Lickers, 2006); however, it is illegal to send tobacco by mail across provincial borders in Canada. In order to resolve this tension, I hand made and sent participants empty tobacco pouches. They expressed appreciation and acceptance of this method.

I follow S. Wilson (2008) in cultivating and maintaining the relationships that are central to conducting Indigenous research. I regularly consult with my family, and spend time with Elders and other traditional knowledge holders, engaging and assisting with cultural events and ceremonies.

By spending regular time on the Land by myself and with family, I also work to reinforce and refresh the bond that I feel to the Land and my conviction to continue research as land-based and land-respectful work, bringing together thought and action in environmental education and research (Scully, 2012).

Reciprocity and Community Accountability

As previously mentioned, both interpretive and Indigenous researchers emphasize reciprocity as an essential component of ethical practice. The concept of reciprocity is informed by questions such as "For whom is this research?" and "Who is benefiting from this research?" (Lemesianou & Grinberg, 2006, p. 230). Reciprocity also means recognizing that, while you may be conducting research with a select group of individuals, they are members of a greater community, and it is important to honour and recognize that community. In Indigenous contexts, this is especially important because certain forms of knowledge (e.g., ceremonial) are communally owned and governed; their use and dissemination is often dictated by strict community protocols and traditions. Researchers, Indigenous and non-Indigenous alike, must ascertain in advance, through community relationships and consultation, what protocols might apply to their proposed investigations (Kovach, 2010; Wilson, S., 2008). For example, if you were working with participants from a specific geographical or cultural community, it would be important to ensure that the research findings are shared publicly in a variety of accessible formats to ensure accuracy, gain approval for its use, and share the findings for the benefit of the greater community (Kovach, 2010).

Kovach (2010) also notes that in more ethnically or geographically diverse contexts where participants come from a variety of cultural and/or geographic backgrounds, it is still important that they are apprised of and benefit from the findings of research. For example, Kovach's doctoral study was with Indigenous academics across Canada, so she shared her results with other Indigenous graduate students and academics at her home university and beyond through publications and presentations.

Public knowledge advocates such as Willinsky (2006) also emphasize the importance of sharing research findings publicly and ensuring that participants benefit in some way. He argues that the evaluation of academic research, especially in the field of education, should be based on its free public knowledge contribution through a variety of media for the benefit of all citizens. Willinsky also notes that "the warrant for conducting research is that such work will contribute to knowledge, which is regarded as a matter of public good" (p. 440).

Methodological Implications

Conscious of past examples of misappropriation and misrepresentation of Indigenous cultures and knowledge by Western researchers (Hermes, 2000; Simpson, 2004), interpretive researchers such as Kincheloe and Steinberg (2008) astutely comment:

> We simultaneously heed the warning that the emerging Western academic interest in indigenous [sic] knowledge may not be a positive movement if such knowledge is viewed as merely another resource to be exploited for the economic benefit of the West. Understanding this admonition, we frame indigenous knowledge not as a resource to be exploited but as a perspective that can help change the consciousness of Western academics and their students while enhancing the ability of such individuals to become valuable allies in the indigenous struggle for justice and self-determination. (p. 152)

Kincheloe and Steinberg's comments seem to match well the perspectives of Indigenous scholars such as Snow (1977/2005) and Cajete (2001) who suggest that the future success of our society will require a combination of the strengths of all cultures. Others warn of the dangers of cultural misappropriation and misrepresentation (Hermes, 2000; Simpson, 2004). In consideration of such warnings, is it possible for someone who is not Métis to enact a methodological métissage such as the one that I have presented?

Building on the work of leading *métisseurs* (Chambers, Donald, & Hasebe-Ludt, 2002; Hasebe-Ludt, Chambers, & Leggo, 2009), my experiences with these two studies suggest that it is possible. It is critically important to note, however, the distinction between a person's identity and their philosophies and practices. Adopting métissage as methodology does not mean misappropriating a Métis identity. It would be inappropriate for a non-Indigenous or non-Métis researcher to simply adopt a métissage approach without first beginning with a bricolage, carefully considering the similarities and differences of the specific culturally rooted methodologies that they are attempting to engage and combine (Simpson, 2002) as well as building relationships and trust with Indigenous communities and peoples (Snively, 2009). As Roth (2008) emphasizes, this involves a process of moving from bricolage (conscious integration) to métissage (unconscious blending), further expanding opportunities for intercultural pedagogical praxis grounded in community-based learning and service, a place-based sense of collective connection to land and culture, and support for Indigenous self-determination.

. 3 .

ARTICULATING A *MÉTIS* WORLDVIEW: EXPLORING THE THIRD SPACE

> England had kings, queens, and jacks. But we had the jokers. We *were* the jokers. Outside the deck, across the ocean, dancing our little jigs of happiness.
>
> — Jessica Grant, Come Thou Tortoise (2009, p. 141)

When considering the contemporary blending and/or integration of Western and Indigenous knowledge and philosophies in Canada, one solution that comes to mind is simply adopting the worldview and practices of the Métis people. However, in this chapter I argue that identifying and adopting *the* "Métis worldview" as a model for contemporary métissage (ecological in this case) would be inappropriate because a singularly identifiable Métis worldview does not exist. While certain similarities in language patterns, spiritual beliefs, and other cultural markers can be identified, the diversity between and within Métis communities and people in Canada is greater than their commonalities.

Nevertheless, what can be most commonly identified is a "Third Space" *mentality* (Richardson, 2004): amenability to incorporating two or more cultures, languages, and spiritual traditions on an individual, community, and regional level. Rather than seeking to reduce and essentialize the vast diversity of the Métis world to an exclusive set of cultural and epistemological characteristics, I demonstrate that what is required is an understanding of the *spirit* of métissage—a Third Space mentality. I believe that it is this spirit or mentality that is required for successful ecological métissage. I also present examples of the Third Space mentality in non-Métis communities in Canada as touchstones for the development of ecological métissage in our contemporary society.

Métis Culture in Canada: Controversial Origins and Confusing Terms

Gibbs (2000) questions whether a singular Métis culture and worldview can actually be defined or if, in fact, a multitude of values and practices exist. The diversity of Métis cultures and epistemologies both in the past and present is so broad that any attempt to distill a singular definition of *the* "Métis worldview" is impossible. While some Métis draw fairly equally on the Aboriginal and European traditions of their ancestors, others identify more firmly with one or the other (Edge & McCallum, 2006). Thus, the dynamic nature of Métis cultures and worldviews in Canada also resists definitive comparisons with their Western or Aboriginal counterparts.

Rather than seeking to provide a definitive description of the specific cultural, linguistic, and spiritual characteristics of Métis peoples, my intention in the following is to elucidate the essence of a Métis *mentality* and how it might contribute to environmental education in Canada.

Who Are the Métis?

"Who are the Métis?" is a seemingly simple question that most likely elicits a host of mental images for the average Canadian: Louis Riel, Gabriel Dumont, the Red River Rebellion, voyageurs, Half-European/Half-Aboriginal, buffalo hunters, the sash, the Red River jig, the list goes on. Many of us have strong preconceptions of the Métis peoples in Canada that are primarily based on Western Métis culture. However, a review of Métis literature, along with my own experiences with Métis peoples across Canada, has revealed to me that the story is much more complex. For example, Métis academic Rivard (2008) suggests:

> As early as the French regime…episodes of *métissage* created *Métis geographies*, that is, new cultural spaces between Indian and European societies, spaces conducive to

> Métis ethnogenesis...these episodes of *métissage* and Métis geographies were distinct from one another because of the different peoples involved and the different contexts in which they took place...The diversity and recurrence of Euro-Indian *métissage* in early Canada make it a valuable topic of investigation. (p. 46)

Foster (1978/2007a) also suggests that:

> Much confusion surrounds the use of the term "Métis". While scholars and laymen [sic] alike agree that the term refers to persons of mixed Indian and Euro-Canadian ancestry, it is difficult to obtain a more precise definition... In essence, such questions are problems in historical understanding. (p. 21)

Like the Trickster figure in Aboriginal mythologies (Graveline, 1998), Métis peoples are hard to pin down; we are shape shifters. Some Métis have blended into Aboriginal communities while others now "pass" as Euro-Canadians (Fujiwara, 2001–2003; Richardson, 2004). Others have maintained or revived their Métis identity and culture, living and viewing the world through a lens that blends both Aboriginal and European perspectives (Richardson, 2004). As is discussed later in this chapter, these kinds of identity dynamics can create tension and confusion within Métis families and communities as well as with non-Métis people attempting to understand or define us.

Métis history and identity are highly complex and controversial subjects. Entire theses and dissertations have explored these topics at length (e.g., Gibbs, 2000; Richardson, 2004). It is not my intention here to replicate such inquiries. Rather, my goal is to explore the complexity of Métis histories and identities in order to articulate the prohibitive difficulty of defining *the* "Métis Worldview" as it relates to ecological philosophy and education.

Origins

A review of literature reveals a broad variety of views on the origins of Métis peoples in Canada. While some argue that the Métis peoples did not crystallize as a cultural group until the conflicts and resistance centered on Red River and Batoche in the mid to late 1800s (e.g., Hanson & Kurtz, 2007), others suggest that Métis communities influenced by both Aboriginal and European cultures, emerged across Canada and the northern United States soon after the onset of European colonization (Gibbs, 2000; Hanrahan, 2000; Karahasan, 2008).

In her doctoral dissertation, Karahasan (2008) presents an exhaustive exploration of métissage in Canadian history. She provides extensive evidence of the emergence of mixed communities in Acadia in the sixteenth century and works through present-day Québec, the northern States, Ontario, Louisiana, and *finally* the northwestern states, provinces, and territories. Métis peoples in all of these areas established unique linguistic and cultural communities representative of both their European and Aboriginal heritage. For example, Métis and non-Métis peoples in the Maritimes commonly used an argot of French, Basque, and Mi'kmaq. Contemporary cities such as Louisbourg, Detroit, Green Bay, and Sault St. Marie, Ontario were also well-known Métis communities during the fur trade. Some of these communities have preserved their Métis consciousness (e.g., Sault St. Marie, Ontario), while others were eventually absorbed into the increasingly dominant French or British colonizing societies. Their inhabitants were likewise absorbed, or, as in the case of my maternal grandfather's family from the Baie des Chaleurs region of northeastern New Brunswick or my paternal grandmother in the southern Great Lakes, scattered across the western United States and Canada.

While an abundance of literature describes Métis communities in western Canada, those of the East are less extensively documented (Karahasan, 2008). This is not to say that they did not or do not exist. For example, Hanrahan (2000) presents an inquiry into the health issues of contemporary Labrador Métis communities. Bartels and Bartels (2005) also provide an account of the "Jackatars", people of mixed French and Mi'kmaq descent in Newfoundland. Gibbs (2000), Karahasan (2008), and Mallet (2010) also relate compelling historical and genetic evidence indicating that many Acadian and Québécois families are actually Métis in origin.

However, due to inconsistent and incomplete record keeping, the prevalence of unsanctioned "country marriages" (especially in the Northwest) (Foster, 1994/2007b), and the reticence of record keepers to acknowledge Aboriginal peoples who married into Euro-Canadian families, accurate estimates of the number of Métis families both historically and presently in Canada remains difficult (Karahasan, 2008). For example, my own family records in New Brunswick simply indicate that one of my Acadian ancestors married "a savage" (C.A. Force, personal communication, April 15, 2010). Adoptions, unregistered births, and the forced imposition of European surnames on Aboriginal peoples further confuse historical records (Karahasan, 2008). For these reasons, many Métis families today rely on a mix of often-incomplete official records along with oral histories to retain or rediscover their genealogies.

Unfortunately, as Richardson (2004) suggests, this complexity often results in significant losses of Métis history and genealogy.

As male European explorers and fur traders ventured further into the interior of North America, many established relations with the Aboriginal women whom they encountered (Foster, 1994/2007b; Karahasan, 2008; Nute, 1987). These relationships took on many forms—some were casual or mutually passionate affairs, others were strategically arranged marriages, and some, unfortunately, were cases of rape (Foster, 1994/2007b; Gruzinski, 2004; Karahasan; Podruchny, 2006).

Angleviel (2008) and Gruzinksi (2004) also remind us that not all instances of cross-cultural affairs result in cultural métissage. Indeed, the offspring of early mixed unions in Canada were born into a variety of circumstances; some Métis children were simply absorbed into the Aboriginal societies of their mothers or, less frequently, the European communities of their fathers. However, others, increasingly over time, were raised under the influence of both sides of their heritage. As European settlement, often linked to the fur trade, expanded across North America, an increasing number of European men settled permanently or semi-permanently with Aboriginal women to raise their Métis children together. Sometimes these families settled into distinctly Aboriginal or European communities. However, distinctly Métis communities also emerged as groups of Métis families created new settlements of their own (Foster 1994/2007b; Karahasan, 2008).

Contemporary Métis Identity

Métis identity has become a politically charged and highly controversial subject (Devine, 2010; Foster, 1978/2007a). This is largely due to the fact that Section 35 of Canada's 1982 Constitution Act recognizes the Métis, along with the Inuit and First Nations ("Status Indians"), as Aboriginal peoples, but it doesn't define who exactly is or isn't Métis. This has left interpretation open to the various regional and federal groups that represent the interests of the Métis peoples. The Federal government's now defunct Aboriginal Portal (2011) was an excellent illustration of this situation; people interested in applying for a Métis status card were directed to a web page containing over a dozen different regional and national organizations. Each of these groups has their own definition of just who is Métis with varying levels of inclusivity. For example, one of several federally recognized organizations, the Métis National Council (2002), defines a Métis person as:

> A person who self-identifies as Métis, is of historic Métis Nation Ancestry, is distinct from other Aboriginal Peoples and is accepted by the Métis Nation.
>
> "Historic Métis Nation" means the Aboriginal people then known as Métis or Half-Breeds who resided in the Historic Métis Nation Homeland;
>
> "Historic Métis Nation Homeland" means the area of land in west central North America used and occupied as the traditional territory of the Métis or Half-Breeds as they were then known;
>
> The Métis Nation's Homeland is based on the traditional territory upon which the Métis people have historically lived and relied upon within west central North America. This territory roughly includes the 3 Prairie provinces (Manitoba, Alberta and Saskatchewan), parts of Ontario, British Columbia and the Northwest Territories, as well as, parts of the northern United States (i.e. North Dakota, Montana). (http://www.metisnation.ca/who/index.html)

This definition is controversial as it excludes a significant number of contemporary Métis in Canada (Richardson, 2004), such as the previously discussed Métis people of the Maritime provinces (Bartels & Bartels, 2005; Hanrahan, 2000) and Québec (Karahasan, 2008). I agree with Karahasan who describes the present state of internally divisive Métis politics as "technocratic genocide" (p. 232). For example, people without quantifiable proof of "historic Métis Nation" ancestry such as scrip records, documentation of controversial compensation given in exchange for title to land that was given to some Métis families (Vizina, 2008), may be excluded from the contemporary Métis Nation status despite generations of familial Métis consciousness, community involvement, and treatment as such by the broader society. These internal divisions can result in serious identity issues for contemporary Métis peoples.

Even the unifying term "Métis" can be questioned—only recently has this term come to universally signify someone of mixed Aboriginal and European ancestry in Canada (Foster, 1978/2007a; Karahasan, 2008). While it has been in use since the beginning of French settlement in Canada and is used, in fact, throughout the French colonies (Karahasan, 2008; Lefèvre, 1989), many other terms were and still are used to describe and identify those peoples who are now commonly known as "Métis" in Canada.

"Métis" is French for "mixed", originating from the Latin "miscere" or "mixtus" (Karahasan, 2008; Nguyen, 2005). Historical sources indicate that the term "Métis" was first used to describe people of mixed Aboriginal and European ancestry as early as the sixteenth century in present-day Canada

(Rivard, 2008). However, while people of mixed European and Aboriginal origins in Canada have been identified as "Métis" throughout history, other French terms such as "bois-brûlé" ("burnt-wood", a reference to skin colour) and "Canadien" were also commonly used (Karahasan, 2008). In the English-speaking world of the Hudson's Bay Company, derogatory terms such as "half-breed" or "half-caste" were employed as well as the less pejorative "Rupert's Lander" or "country born" (Foster 1978/2007). Terms for Métis peoples can also be found in Cree, Ojibwe, and Michif (a Western Métis language). In Ojibwe, the term *Wissakodewinmi* means "half-burnt woodmen" (Gibbs, 2000). Another Ojibwe term commonly used is *Wesahkotewenowak*, which roughly means "fresh plant shoots that grow after a fire" (Edge & McCallum, 2006), while in Cree, *Otipemisiwak*, meaning "free men" or "people who own themselves", is common (Edge & McCallum, 2006; Karahasan, 2008; Vizina, 2008), referring to the independent lifestyles of early Métis traders. In the Cree-Michif dialect, *Apihtaw'kosisan* is commonly used and means "half-son" (Edge & McCallum). Hanrahan (2000) also reports that the terms "settler", "breeds", and "esquimaux" were used in the past in reference to the Métis peoples of Labrador. In Newfoundland the derogative term "Jackatar" described those of mixed European and Aboriginal ancestry (Bartels & Bartels, 2005). Most recently, for simplicity and political unification, the term "Métis" has been adopted across Canada (Foster, 1978/2007; Karahasan, 2008). This regional and linguistic diversity is important to recognize because it demonstrates that people of mixed Aboriginal and European ancestry have been recognized as distinct peoples across Canada throughout history, casting further doubt on the existence of a singular Métis culture.

Based on the complexity of Métis history and identity, Richardson (2004) provides a broad definition of contemporary Métis in Canada as those people of mixed Aboriginal and European (and possibly other) origins who self-identify as "Métis" and who incorporate aspects of Aboriginal and European cultures and worldviews into their identity and lifestyle.

Fujiwara (2001–2003) suggests that many people of mixed Aboriginal and European origins identify exclusively as either "Aboriginal" or "White". Despite being of mixed origins, Métis peoples may choose to identify, for various personal, political, or physical reasons with only one side of their ancestry. Kienetz (1983) proposes that, for many people, identifying as White, Aboriginal, or Métis is simply a matter of philosophy. Gibbs (2000) also states:

> There is still little recognition by the dominant society that Métis identity is not static but dynamic, and that there is always a philosophical element in the question of any identity. (p. 66)

This kind of identity diversity is common even within a single family (Richardson, 2004). The reflections of Métis scholars such as Adams (1999) and Richardson (2004) who describe identity conflicts within their own families resonate with my own experiences; some of my family members, myself included, self-identify as "Métis", while others call themselves "White" and are reticent to discuss our Aboriginal origins, despite being visibly "Aboriginal" to other people.

Bonniol and Benoist (1994) and Richardson (2004) also suggest that how a person of mixed heritage self-identifies or is identified by society is often related to their physical appearance, a highly variable factor in mixed families. However, this variability can also create turmoil for people who are identified by society as fitting into one cultural group when they actually identify with another, or perhaps several others. This kind of identity conflict often leads to what Cajete (2001) calls the "split head" phenomenon—a feeling of being split between two or more worlds without fitting anywhere.

Foster (1978/2007a) relates that, historically, government categorization of who was Métis was highly subjective. He suggests that being identified as Métis was sometimes based on lifestyle rather than ethnicity. For example, several Iroquois fur traders were granted Métis scrip in the prairies based on their semi-nomadic Métis-like lifestyles. Karahasan (2008) also provides evidence that some people who were considered "Métis" were actually Europeans who had adopted Aboriginal ways. Gibbs (2000) and Foster both argue that it sometimes simply came down to who was or wasn't physically there to be counted when government agents came to visit.

Peoples of mixed Euro-Canadian and Aboriginal ancestry were and are present across Canada. Karahasan (2008) suggests that several Métis cultures formed "synchronically" at different places in Canada in many different forms. She states:

> The several Metis groups reflect a diversity of Metis experiences, and show that Metis developed, in fact, a proper identity that differed from both Europeans and Indians, while at the same time showing characteristics that could be just as typical of Europeans (settlement and agriculture) as of Indians (nomadism and chase). (p. 202)

As Karahasan relates, Métis peoples formed a broad new variety of cultures based on the geographical and cultural characteristics of each region drawing on the diverse cultures of both European and Indigenous ancestors. This resulted in a plethora of Métis cultures across Canada. For example, a family of French-Dené Métis in the present day Northwest Territories might bear minimal cultural or linguistic similarity to a Scottish-Inuit Métis family in Labrador. However, in my opinion, both of these families have the right to self-identify as "Métis". Gibbs (2000) provides evidence from a 1981 national census indicating that self-identifying Métis were present in every province. She emphasizes the regional differences between Métis communities and suggests that this lack of a collective identity and ideology results in self-doubting and identity issues for many contemporary Métis. Gibbs also states:

> A section of the public's perception is that the Métis seem unable to forge a sense of solidarity and identity but that is not strictly true... The tremendous differences that far outweigh the commonalities existing between the various Métis communities do not appear to lessen family ties but rather appear to strengthen that sense of solidarity and identity...Strangely, First Nations people are not expected to conform to one rigid image of "Nativeness" but that is not held as true for the Métis. The Métis, of course, have a commonality of Aboriginal ancestry but they do not have a clearly defined image of that source of identity, and therefore lack a compliance ideology on which to build a collective identity. The question of collective Métis identity needs to be clarified so that the Métis are perceived by the public not only as a political entity but also as a unique Aboriginal society. (pp. 59–60)

Gibbs' comments are reminiscent of similar critiques of "pan-Indianism"—applying overly broad generalizations to First Nations peoples—that has been consistently challenged on the grounds that, while First Nations cultures across North America are related and *some* commonalities are evident in worldviews, customs, and practices, there is also an enormous amount of diversity that resists overgeneralizing (Rosser, 2006; Waldram, 2000). Gibbs' reflections also remind me of Richardson's (2004) nod to Hemingway when she comments that, for contemporary Métis people who don't have regular contact with other Métis, opportunities for interaction with others to whom we can deeply relate are a "moveable feast". Considering these kinds of statements led me to question whether métissage is unique to Canada or are there similar cultures in other areas of the world?

Global Voices: International Perspectives on Métissage

While famous Métis scholars such as Campbell (1983) and Adams (1999) have drawn attention to the contemporary experiences of Métis peoples in Canada, voices from around the world also add to contemporary Métis discussion. A review of literature reveals that there are actually Métis peoples across the globe; wherever the French colonized, the term "Métis" was (and often still is) used to describe peoples of mixed European and Indigenous ancestry (Nguyen, 2005). I was surprised and delighted to discover the similarity of experiences and issues discussed by Métis peoples worldwide.

I find a special resonance with several scholars from French Polynesia. Their descriptions of the importance of the canoe, or *wa'ka*, in Indigenous Polynesian cultures (Keown, 2008; Ramsay, 2008; Wilson, N., 2008) remind me of its similar historical and contemporary place in Canadian culture. In a manner similar to canoes here in North America, Ramsay (2008), Keown (2008), and Mateata-Allain (2008) relate how the wa'ka served as both a literal and physical vehicle of métissage in Polynesia by carrying people across the ocean to meet, share stories, and mix culturally and biologically. Mateata-Allain (2008) comments further that modern colonial political and linguistic boundaries have restricted traditional intercultural exchange in Polynesia—a phenomenon also common to North America.

Several Métis voices also emerge from Asia. Stoler (1992) describes métissage in Southeast Asia from a personal perspective as both an advantage and disadvantage. Stoler suggests that while being Métis sometimes helps you to fit into certain situations, it may also exclude you from others. Choo (2007) and Nguyen (2005) also discuss métissage in Southeast Asia and the role of Métis or "Eurasian" peoples as intercultural mediators—people who are able to navigate both Asian and European society. Echoing the previously discussed debate around Métis identity in Canada, Choo also raises the issue of who "qualifies" as Métis in Southeast Asia: Only historically mixed families or contemporary ones as well? Choo also relates that some young Eurasians in Southeast Asia try to hide their European roots, preferring to blend into mainstream society.

In Africa, "Métis" is also used to describe peoples of mixed African and European ancestry. For example, in the Congo, Belgian men often engaged in sexual relations with Congolese women; their children were called "Métis"

(Rahier, 2003). Rahier, a Congolese Métis himself, reports that there was initially societal and institutional confusion regarding how Métis should be treated: were they Black or White? However, as the Métis population grew, "this ambiguity came to characterize the lives of the growing population of métis" (Rahier, 2003, p. 86).

The term "Métis" is also used in French-speaking areas of the Caribbean such as Haiti and Guadalupe to describe people of mixed European and African ancestry (Niort, 2007). Niort relates that, as in many other colonial nations, European men and African women, who were usually slaves, often entered into sexual, but rarely marital, relationships. Similar to Africa, Caribbean societies such as Haiti were initially uncertain how to categorize the Métis offspring of these relationships, but as their numbers grew, they came to represent an elite upper class of servants, acting as intermediaries between the ruling Europeans and enslaved Africans.

I find it interesting to see the links and commonalities among Métis communities around the world. I also find it heartening to see the broad use of the term "Métis" which, as was previously discussed, has become highly politicized here in Canada (Karahasan, 2008). There are many intriguing parallels between Metis communities around the world with issues such as old vs. new families, denial of heritage, and links to ecological identity.

An exploration of Métis history and contemporary issues in Canada and internationally may leave us with more questions than definitive answers. However, an ambiguous result may be, in itself, an answer. Keeping this in mind, rather than providing definitive definitions and comparisons, the following section attempts to clarify the essence of a Métis or Third Space mentality as it relates to the present inquiry.

Exploring the Third Space: Understanding a Métis Worldview

In her doctoral dissertation, Métis scholar Richardson (2004) adapts Bhabha's (1998) "Third Space" concept—an intercultural existential territory resulting from cultural blending—to describe a Métis mentality. Richardson compares the Métis Third Space to the "First Space" of the dominant Euro-Canadian society and the "Second Space" of colonially subjugated Aboriginal peoples. Richardson's use of these terms is an invitation for us to further explore the Third Space in a Métis context.

During a recent conference presentation (Lowan, 2010), an audience member astutely suggested to me that the First Space here on Turtle Island (North America) was, in fact, Aboriginal, followed by the Second Space introduced by the colonizing Europeans, which resulted in the Third Space of the Métis. This revised interpretation of Richardson's (2004) Third Space is presented in the figure below.

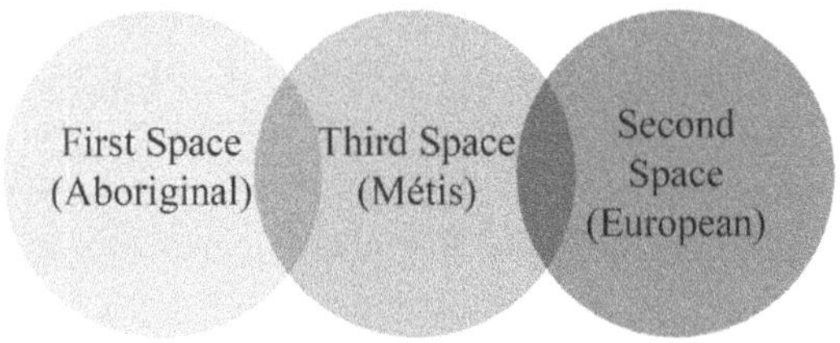

Figure 2. The Third Space (Lowan, 2011a, 2011b).

The Third Space is a place where Western, Aboriginal, and possibly other beliefs, philosophies, values, and knowledge intersect, co-habit, and intermingle (Richardson, 2004). Zembylas and Avraamidou (2008) propose that the Third Space represents "points of departure" (p. 990) to challenge existing norms—a "landscape of becoming" (p. 990) that stimulates "transformative practice" (p. 990).

This discussion focuses on the Third Space between Western and Aboriginal cultures but it need not. For example, Chambers (2002) discusses the métissage within her own Euro-Canadian family. She reminds us that, not too long ago, even being of mixed English and Scottish heritage was notable in Canada, as both groups were still very conscious of their distinct origins.

The following illustrative examples of the Third Space are drawn from examples of Métis lives and philosophies. What might we learn from the lifestyles and communities of Métis peoples in the past and present? How might specific examples of Métis life and culture in Canada inform our understanding of the Third Space? Are there examples of non-Métis people embracing a Third Space mentality? Further discussion of the Third Space and other related concepts will be continued in the following chapter.

Interpreting Métis Lifestyles

Karahasan (2008) suggests that métissage is a concept that needs to be interpreted rather than defined. The following section presents information related to the historical lifestyles and practices of Métis peoples across Canada.

I will elucidate the Third Space mentality through interpretation of various examples of Métis cultures and communities as they are manifested in spirituality, marriage, education, and geography. What might these practices and traditions tell us about a Métis worldview or Third Space mentality?

Spirituality

Historically, Métis communities were spiritually diverse (Foster, 1994/2007b; Fujiwara, 2001–2003; Vizina, 2008). Some followed Anglican or Catholic practices (Foster, 1994/2007a), while others also incorporated Aboriginal traditions such as the Sweatlodge[1] (Edge & McCallum, 2006). Fujiwara (2001–2003) and Duval (2001–2003) report that many communities followed a kind of "folk Catholicism", blending both Catholic and Aboriginal spiritual practices. Métis scholar Vizina (2008) adds that:

> The spirit of the Métis and the spiritual practices of the Métis are as complex as the ancestral roots of their Indian and European culture and languages...[Métis] were comfortable blending them together. (p. 175)

Karahasan (2008) reports that when the first Catholic Mission was established in the Red River settlement in 1818, the missionaries found the Métis practicing a broad and blended variety of syncretic folk religion ranging from purely Indigenous traditions to those of the Catholic Church. Unfortunately, due to the predominantly oral tradition of Métis cultures, written records of how this blend was embodied in practice are limited, indicating an area in need of further research (Dorion & Préfontaine, 1999).

Duval (2001–2003) shares that many Métis were reluctant to contribute funds to build the first Catholic Church in Red River, happy to continue with their less formal religious practices. However, some Red River Métis such as Erasmus (1999), a Protestant, devoted much of their lives to missionary work with Aboriginal peoples.

Even once the Catholic Church was established in the Red River Valley, they were still frustrated by the Catholic Métis' openness to intermarriage with their non-Christian Aboriginal and Protestant neighbours (Duval, 2001–2003), further demonstrating the Métis mentality that led to increasing European and Aboriginal métissage. Duval also suggests that the Roman Catholic Church tried to curtail the Métis' semi-nomadic traditions by discouraging hunting and fishing and encouraging them to focus full-time on settled agricultural pursuits. Apparently they were largely unsuccessful in this project, but did succeed in temporarily banning hunting and fishing on

Sundays. However, Duval reports that the missionaries eventually conceded and began accompanying their Métis parishioners on hunting expeditions.

Métis scholars Edge and McCallum (2006) provide further insight into the contemporary spiritual views of Métis Elders in the prairies. They report that, while a strong sense of spirituality and faith exists in the Métis community, there is no one view common to all western Métis. For example, while some Elders attend and facilitate Indigenous ceremonies such as the Sweatlodge, pipe ceremonies, and the offering of tobacco, others live a predominantly Christian existence. However, many embrace both, drawing on Christian and Aboriginal spiritual traditions to create their own unique métissage. Reflecting an awareness and acceptance of this complexity, Métis Elder McCallum (Edge & McCallum, 2006) suggests that we should all reflect on our own spirituality and decide for ourselves which aspects of our Aboriginal or European roots to incorporate.

Marriage

What might early Métis marriage customs tell us about their worldviews? Several historical sources provide accounts of marriage *à la façon du pays* (Foster, 1978/2007a; Karahasan, 2008; Podruchny, 2006). These "morganatic" or unsanctioned "country marriages" were often a blend of European and Aboriginal practices, unsanctioned by the Church and variously incorporating rituals and traditions of both cultures. For example, European and Euro-Canadian men who wished to marry an Aboriginal woman often followed local customs of paying a bride-price (e.g., a horse) and living with the bride's family until the birth of their first child (Foster, 1994/2007b). This practice allowed the groom to learn local knowledge and skills under the watchful eyes of the bride's family.

Métis marriage ceremonies were often quite simple and regularly employed a blend of Western and Aboriginal traditions. Historian Podruchny (2006) relates:

> The wedding custom was a blend of Aboriginal and French Canadian practices… Some marriages included rituals such as the smoking of a calumet [pipe] and a public lecture by Aboriginal elders on the duty of a wife and mother. The new Aboriginal bride was then cleaned by other women at the [fur] post and then clothed in "Canadian fashion", which consisted of a shirt, short gown, petticoat, and leggings. (p. 271)

As with many other aspects of Métis life, early marriage customs reflect flexibility on both sides to incorporating Aboriginal and Western practices.

This flexibility was retained and variously interpreted as Métis communities grew throughout Canada.

Healthy Land, Healthy People

> Territoriality played a vital role in forming Metis identity. The attachment to indigenous soil was one of the characteristics of Metis community formation. (Karahasan, 2008, p. 211)

What might accounts of early Métis relationships to and interaction with the Land tell us about their worldviews? Similar to other Aboriginal cultures (Cajete, 1994; Simpson, 2002) as well as certain Western schools of philosophy such as deep ecology (Naess and Rothenberg, 1990), a deep connection to the Land is a prominent feature of Métis cultures in Canada (Edge & McCallum 2006; Gibbs, 2000; Hanrahan, 2000).

Gibbs relates a western Métis view that the Land is naturally perfect and thus people should adapt themselves to it, not the reverse. Hanrahan seems to concur when she reports the Labrador Métis perspective that, rather than people taking care of the Earth (as in a Western stewardship model), the Earth takes care of the people. This means that it is the people's responsibility to act with humility and care towards the Earth and all creatures.

Echoing the perspectives of Aboriginal scholars such as Cajete (1994), Hanrahan (2000) shares the views of Labrador Métis Elders who believe that human health is intimately linked to the Land. They suggest the Land and the Sea are the basis for a deep interconnection within and between individuals and their human and non-human community. This view leads them to maintain practices such as eating wild foods harvested from the Land as well as following traditional Inuit and Mi'kmaq seasonal rituals such as preventive "spring cleaning": fasting and cleansing oneself using traditional plant medicines at the end of winter. Many Labrador Métis Elders continue to use traditional plant knowledge in conjunction with Western medicines (Hanrahan, 2000).

These kinds of perspectives can also be found in the Western world; Euro-American philosopher and farmer Berry (2009) describes the traditions of Euro-North American farmers who not only grew crops and raised livestock, but also went hunting and fishing on a regular basis. These farmers were keenly aware of the complex relationships between their own cultivated land and the surrounding "wilds".

Edge and McCallum (2006) report similar perspectives from western Métis Elders. At a recent Elders' gathering various beliefs were shared on the importance of the Land in Métis cultures. Similar to the Labrador Métis, western Métis Elders shared that a connection to the Land is a key factor in Métis health, knowledge, wisdom, and stories. Many Elders also stated that, as in many North American Aboriginal cultures, the concept of "All My Relations", recognizing the deep interconnectedness of all things, is foundational to a Métis worldview (Edge & McCallum, 2006). Edge and McCallum also reported dual approaches to Métis health and healing that incorporate both Western and Aboriginal traditions. They emphasize the diversity in the Métis community, with Elders falling on a spectrum from Aboriginal to Western views and practices.

Settlement and Housing

Given the fundamental importance of the Land in Métis cultures, how were Métis worldviews represented in early settlement and housing practices? Gibbs (2000) reports that many prairie Métis families and groups were semi-nomadic—maintaining small farms, but also travelling across the prairies to hunt and trade on a regular basis. Karahasan (2008) suggests that with increased missionary and European settlement, the semi-nomadic lifestyles of many Métis resulted in hierarchies forming in communities such as Red River in Manitoba. She proposes that the degree of a family's settlement reflected their status in the community; fully settled Europeans and Euro-Canadians were at the top with the more settled Anglophone (often Scottish) Métis next followed by the buffalo-hunting Francophone Métis and finally the surrounding Aboriginal groups who were seasonally nomadic.

Troupe (2009) reports that many western Métis settlements such as Batoche in Saskatchewan had a "big family village feel" (p. 60). Troupe and Podruchny (2006) both note that many Métis communities across the West followed French Canadian settlement patterns using a river-lot system where each family was allotted a narrow strip of land that provided universal access to a river or lake. This system contrasts with the American-style land settlement pattern of dividing land into wider sections that was autocratically introduced by the Canadian government when it took over management of Rupert's Land in 1869 (Vizina, 2008). This unilateral action reduced regular water access to those landholders bordering the water source, a less egalitarian approach (Troupe, 2009).

Métis homes often reflected both Aboriginal and Western roots. Burley and Horsfall (1989) state that western Métis homes often appeared European from the outside, but were very open inside, reflecting Aboriginal values of extended family and community. Karahasan (2008) relates a derogatory description by one European observer of Malecite Métis homes in St. Malo, Québec as European style homes inhabited by people living in "Indian style". Similar to other cultural aspects, early Métis settlement and building patterns reflected a Third Space mentality that incorporated both Aboriginal and European traditions and practices.

Education

Many Métis families educated their children in a blended European and Aboriginal style in various forms depending on the time period, community, or individual family. For example, Métis boys from wealthier elite families in the Red River region, such as Louis Riel, were often sent east in their teens to Montreal (Duval, 2001–2003) or even Europe (Brown, 1983) for more formal educational training. Most records indicate this was less common for Métis girls who tended to stay at home to learn from their mothers and boys from poorer families who were more likely to remain in the Northwest developing their hunting, fishing, and agricultural skills (Brown). Brown also notes that many Métis children, girls and boys alike, were also exposed to a combination of Western-style schooling in the Northwest along with traditional Indigenous knowledge that they learned experientially on the Land with their families.

Edge and McCallum (2006) also report that both Indigenous and Western modes of education were and still are valued by many Métis communities. They indicate that along with Western style approaches, experiential, land-based educational techniques were and still are employed with Métis youth. Edge and McCallum suggest that aspects such as self-esteem, self-reliance, responsibility, and traditional wilderness skills are best taught on the Land and they stress the importance of taking youth outside on a regular basis. They emphasize that land skills are very important.

Hanrahan (2000) notes similar values in the Labrador Métis communities that she researched. Echoing Edge and McCallum (2006), other Indigenous scholars such as Simpson (2002) and Pashagumskum (2014), and the findings of my own master's research (Lowan, 2008; 2009), Hanrahan suggests that Labrador Métis Elders believe that values such as self-reliance, independence, and adaptability are best taught through experiences on the Land.

Edge and McCallum (2006) stress the importance of helping Métis youth develop their sense of personal and cultural identity and belonging. McCallum suggests that they need to learn to be "bi-cultural in a multicultural world" (p. 113), taking the best of all the cultures to which they are exposed and blending them together, rather than keeping them separate. McCallum's comments strongly remind me of the concept of Two-Eyed Seeing promoted by Mi'kmaq Elder Marshall (Bartlett, 2005; Hatcher, Bartlett, Marshall & Marshall, 2009), a concept that will be explored in the following chapter.

Language

The Michif language arose in various forms in many western Métis communities from the intentional blending of French with Aboriginal languages (Bakker & Papen, 1997; Crawford, 1985; Rosen, 2003). Crawford suggests that the consciously constructed balance of French nouns and Aboriginal verbs in the structure of Michif is reflective of the values of the Métis societies that equally valued their European and Aboriginal roots. Saskatchewan Métis scholar Vizina (2008) also states:

> Until recent times, most Métis spoke multiple languages and many were literate in French or English. The Michif language was a unique outcome...of Métis mixed ancestry and creativity. Just as the grammar and lexicon of Michif is unique, the stories of the Métis also combine elements, perspectives and traditions of their ancestral lineages. (p. 175)

I agree with Métis Elder McCallum who relates Métis diversity to the linguistic diversity *within* Michif (Edge & McCallum, 2006). In reference to the revitalization of Michif, he states:

> There is this whole idea [of Michif] that they are promoting right now. It comes from Manitoba and it is mixed with Ojibwe. It has a "zheh" sound. We don't have that sound back home. We have a different dialect of what you call Michif or Cree. And there is probably not as much French, as those French Michif...I've heard that they are trying to standardize it, to make it one. And, that is kind of offensive to the people who don't speak that language...it offends me anyway...I like to speak my language, the language of my people, from where I come from. (pp. 101–102).

The previous section has outlined various aspects of Métis lives and traditions that reflect a blended Third Space worldview. In areas such as religion, marriage, education, and language, Métis peoples incorporated aspects of their European and Aboriginal roots in various ways. One might be tempted to consider simply adopting customary Métis cultural practices and traditions as an easy way to achieve contemporary métissage. However, as stated at the beginning of this book, I disagree with that approach. As discussed in previous sections, Métis peoples and cultures across Canada are very diverse and, as such, resist broad generalizations and comparisons. While it is true that many more Canadians than is collectively acknowledged are actually Métis (Karahasan, 2008), it would still be inappropriate for the non-Métis Canadian majority to simply appropriate Métis cultures and practices in a neocolonial (Zembylas & Avraamidou, 2008) manner. What then, would be a more appropriate approach? Are there any examples of non-Métis people in Canadian history adopting a Third Space mentality?

Exploring the Third Space in Non-Métis Communities

The following section presents three examples of the Third Space in predominantly non-Métis communities—the early French voyageurs, Chinook Wawa, and contemporary Aboriginal education. What might we learn from these examples in our search for strategies to guide the development of ecological métissage?

Voyageur Worldviews

The voyageurs are often associated with the emergence of Métis communities in northwestern Canada. Their well-documented narrative tells us that primarily French Canadian voyageurs under the banners of Montreal-based ventures such as the Northwest Company ventured progressively further from their *habitant* homesteads along the St. Lawrence River, eventually spending winters in the Northwest as *hivernants en dérouine* where they increasingly intermingled and settled with Aboriginal women to produce significant numbers of Métis offspring, many of whom formed identifiable and unique communities of their own (Foster, 1994/2007b, Karahasan, 2008). We are told that voyageurs and Aboriginal women formed mutually beneficial relationships for various reasons: material necessity, physical survival, strategic trading alliances, and simply for love (Foster, 1994/2007b).

What is not as clearly documented is the inner world of the voyageurs (Podruchny, 2006). Due to their almost universal illiteracy and the rigidly hierarchical structure of the fur trade where primarily English and Lowland Scottish gentlemen occupied the bourgeois positions of managerial power over the predominantly French (and later Iroquois, Métis, etc.) voyageurs and Highland Scottish "boatmen" (Raffan, 2008), scant evidence remains to tell us stories from their own perspectives. What remains are the accounts of the bourgeois or visitors to Canada who recorded their observations of the habits and customs of the voyageurs.

Podruchny (2006) delves deeper into the voyageur psyche than previous historians. Rather than simply repeating the standard narrative, traditions, and habits of the voyageurs, Podruchny provides an unparalleled *interpretation* of the worldview of the voyageurs. An exploration of her work is relevant to the present task because I believe that insight into the beliefs of the voyageurs provides us with a well-articulated example of a Third Space mentality from a predominantly non-Métis perspective.

Podruchny (2006) suggests that a closer examination of voyageur rituals and habits reveals deep intentionality. Ritual acts such as performing baptisms at every height of land were not simply maintained in order to procure an extra swig of rum from the bourgeois; they served to indoctrinate new voyageurs into the world of their more experienced compatriots, teaching them the rules and values of their trade. Karahasan (2008) suggests that the French colonial project in North America resulted in the unexpected outcome of French people such as the voyageurs adopting Indigenous habits and beliefs as much as the reverse, contrary to the hopes of the French government.

Podruchny (2006) argues that as the voyageurs ventured increasingly further from St. Anne's Church at Lachine, their final stop before leaving Montreal for the Northwest, their world became increasingly Indigenized. Local customs and languages were adopted away from the watchful eye of the Catholic Church. Podruchny and others (e.g., Nute, 1987) note that voyageur rituals often involved a mix of Christian and Aboriginal spiritual practices. For example, a Catholic prayer for safekeeping before a dangerous lake crossing might have been accompanied by the offering of tobacco and a small coin (Podruchny). The voyageur baptism was also a mixed affair that was performed by sprinkling water over a neophyte's head using a sprig of cedar, a sacred plant in many Aboriginal societies. Podruchny also highlights that several of the important sites of voyageur rituals such as Oiseau Rock on the Ottawa River are also sacred Aboriginal spiritual places. As she suggests:

> On a symbolic level, mimicking the location of an Aboriginal spiritual site may have been another attempt at indigenizing themselves or perhaps even garnering spiritual power and protection from Aboriginal forces to bolster the protection of their Catholic saints. (p. 63)

Podruchny also proposes that the voyageurs' habitual smoking of tobacco in their pipes might have been considered as a spiritual act by some, following the beliefs of Aboriginal peoples. She also suggests that some voyageurs adopted Aboriginal spiritual practices to appease their Aboriginal hosts in tense situations as well out of desperation during times of danger or starvation. Podruchny relates that many voyageurs also learned Aboriginal stories (often involving the Trickster) and rituals and incorporated them into their own beliefs and values. For example, some voyageurs offered tobacco and other small items of value to the water at the beginning of every day to pray for safe passage.

Voyageur dress also reflected a blend of Western and Aboriginal styles—often mixing cotton shirts and wool ceintures flechées (sashes) with leather moccasins and beaded tobacco pouches (Karahasan, 2008; Nute, 1987). Not only was their clothing functional, it was also representative of their absorption of both cultures.

Voyageurs who learned Aboriginal languages most likely gained a deeper understanding of Aboriginal cultures (Podruchny, 2006). Indigenous authors such as Simpson (2002) and Cajete (1994) remind us that Indigenous languages are highly representative of the values of their host cultures. As such, a voyageur that learned an Indigenous language would have had greater insight into the culture of its speakers. Evidence of such linguistic interactions are evident in the parallel naming in Aboriginal languages and French of key points along fur trade routes and the eventual development of the Métis language, Michif (see Crawford, 1985; MacDougall, 2006; Rosen, 2003) that I described earlier.

These insights into voyageurs' rituals and beliefs demonstrate the general amenability of the voyageurs to Aboriginal values and ways of life. Many voyageurs took on aspects of Aboriginal cultures and worldviews while maintaining some of their own predominantly French Catholic practices. Podruchny (2006) also proposes that the voyageurs' maintenance of this unique blend was also a way for them to distinguish themselves from the primarily Anglo-Protestant bourgeois.

One could argue that some of the voyageurs adopted Aboriginal spiritual and cultural practices merely to survive in a somewhat hostile physical and cultural environment. This is most likely true, however we know that, while some eventually returned home to their farms along the St. Lawrence, many voyageurs chose to remain in the Northwest, adopting highly Indigenous lifestyles and raising Métis children with their Aboriginal wives (Foster, 1978/2007a; Brown, 1983). The early voyageurs provide us with a well-documented example of a non-Métis people (at least initially) who developed a Third Space mentality in order to survive and thrive under challenging physical, emotional, and spiritual circumstances.

Chinook Wawa

Another example of the Third Space in a non-Métis context is Chinook Wawa, a contact language incorporating various Aboriginal languages, French, English, Hawaiian, Chinese, and Japanese that was commonly spoken socially, in business and government, and written throughout the Pacific Northwest until the 1930s (Backhouse, 2008; Klassen, 2006; Silverstein, 1997). Chinook Wawa is a strong example of a collective Third Space mentality in a non-Métis setting. While the Michif language was generated by peoples of similar cultural origins, Chinook Wawa was developed and adopted by a culturally disparate community (Backhouse, 2008). Some scholars (e.g., Crawford, 1985) emphasize the uniqueness of the Michif language because it is quite rare amongst colonial contact languages as a truly mixed language representing both Indigenous and European languages in vocabulary and grammar. However, Backhouse (2008) observes that Chinook Wawa is also remarkable because it was created and used by a culturally diverse community with the intention of forging a common language available to all.

A contact language such as Chinook represents the Third Space mentality in a different way than a mixed language such as Michif that is primarily spoken by peoples of common cultural origins. I do not mean to suggest that no cultural prejudice or hierarchy existed in the Pacific Northwest while Chinook was in use, but I do find it remarkable that a geographically and culturally disparate community would collectively engage in the project of communal sense-making represented by Chinook Wawa. Rather than forcing non-English or non-French speakers into speaking a completely foreign colonial tongue, the early colonists, other settlers, and the Indigenous peoples of British Columbia and other Pacific regions created and employed a language that used words from several languages—Indigenous, European, and Asian alike. If we consider theorists such as Cajete (1994) and Simpson (2002) who emphasize the strong link between languages and worldviews, then we must conclude that the formation of a language such as Chinook Wawa is significant. It arguably represents a locally generated worldview based on consideration and respect for each contributing culture's values and traditions. Roth (2008) suggests that it is exactly this kind of common jargon mentality that is required to collectively move into the Third Space.

As the colonial project progressed, permanent European settlement increased and more European women were encouraged to emigrate to Canada; the vital role of Aboriginal peoples and intermarriage with Europeans decreased substantially in the late nineteenth and early twentieth centuries (Saul, 2008). However, the Third Space continues to be represented in various ways in Métis and non-Métis communities in Canada. For example, a recent study by Statistics Canada reports that "mixed unions" between a person who is a visible minority and another who is not increased by 33% between 2001 and 2006 in Canada (Milan, Maheux, & Chui, 2010). As mentioned earlier, Statistics Canada also predicts that by 2031, a third of Canadians will be "visible minorities" (Malenfant, Lebel, & Martel, 2010).

Are we in the midst of a renewed expansion of the Third Space where Indigenous and other cultural traditions will once again play a strong role in the framework of our society or will Western norms continue to dominate? Pieterse (1996) suggests that seriously considering cultural hybridity involves the decolonization of the mind. Little Bear (2000b) also comments:

> Colonization created a fragmentary worldview among Aboriginal peoples. By force, terror, and educational policy, it attempted to destroy the Aboriginal worldview—it failed. Instead, colonization left a heritage of jagged worldviews among Indigenous peoples...Yet all colonial people, both the colonizer and the colonized, have shared or collective views of the world embedded in their languages, stories, or narratives. It is shared among a family or group. However, this shared worldview is always contested, and this paradox is part of what it means to be colonized. (pp. 84–85)

Rather than being dominated by a mono-cultural colonial mentality, a Métis worldview or Third Space mentality is open to the perspectives and traditions of several cultures simultaneously. Examples of métissage from our collective past provide inspiring models of intercultural collaboration that transcends these "jagged worldviews". How might we draw on these examples to nurture the Third Space in our contemporary society in areas such as science and environmental education? The following chapter explores current theories and examples of practice that attempt to integrate or blend Western and Indigenous knowledge and philosophies.

Note

1. While many variations exist, the Sweatlodge ceremony is a healing ceremony that is a part of the spiritual life of many First Nations. The ceremony involves praying and singing in a dome-shaped structure that represents the womb of Mother Earth. Hot stones, also called the Grandfathers and Grandmothers, from a sacred fire, heat the lodge (Anishnawbe Health Toronto, 2008; Portman & Garrett, 2006).

. 4 .

IN SEARCH OF COMMON GROUND: TO BLEND OR NOT TO BLEND?

An increasing number of scholars, both Indigenous and non-Indigenous, are asking questions such as, "Is it possible to blend Western, Indigenous North American, and other ecological philosophies and knowledge? Or is it better to keep them separate, but search for commonalities?" Some, such as Cajete (2001) and Snow (1977/2005), suggest that the collective survival of our society will require the *combined* wisdom of Aboriginal and non-Aboriginal cultures.

Renowned Tewa scholar Cajete (2001) relates the story of a female relative who has a "split head"; she is of mixed Euro-American and Tewa ancestry and often feels split between the two cultures. Cajete suggests that many contemporary Aboriginal *and* non-Aboriginal people also have a split head—torn between various cultural and sub-cultural influences and values. He proposes that the ultimate task at hand is to find ways to heal the split head of our collective society, blending the best of Western (and other cultures) and Indigenous cultures to create a unified whole.

Turner (2005), a well-known Euro-Canadian ethnobotanist who has built strong relationships with Indigenous communities on the West Coast, also states:

> Despite the tremendous scientific and technological advances we have made since the Industrial Revolution, humans have not successfully protected our environments or cared for the Earth's other species. Much of today's environmental damage is a direct result of poorly considered use of technology and the impacts of this technological mindset. Our scientific sophistication has not been matched by our caring for the Earth—our environmental ethic...I believe that there are many ideas and approaches we can look to [in the Indigenous world] to help us in our search for better, less harmful ways to live, while maintaining healthy, fulfilling and satisfying lives. (pp. 1–2)

Partnerships between Indigenous and non-Indigenous people and organizations to combat environmental challenges are also growing (Simpson, 2010); however relatively few comprehensive accounts of these initiatives are available (Davis, 2010). As this area of inquiry and practice grows, key issues of concern and debate are rising in the literature. These topics include questions such as how Indigenous and non-Indigenous people can respectfully engage with each other and whether blending or integration can actually be achieved in a Western framework without "watering down" Indigenous knowledge. There is also considerable debate surrounding questions such as: Are Indigenous knowledge and philosophies of Nature a form of science or should they be considered a separate body of knowledge? Also, what are the similarities and differences between Traditional Ecological Knowledge (TEK) and Western science? These kinds of questions also relate to broader discussions of contemporary trends such as globalization and multiculturalism.

The following provides an overview of the key characteristics of and contemporary issues surrounding contemporary discussions of blending and/or integrating Indigenous and Western knowledge and philosophies of Nature. Perspectives and theories from Indigenous and non-Indigenous scholars in North America and around the world are explored and supported with illustrative examples of contemporary practice.

Transcultural Perspectives on Intercultural Engagement

Bricolage, métissage, creolization, the Third Space, integration, hybridity, diaspora, transcultural, intercultural, multicultural, and bicultural, among others, are all terms that one encounters when examining contemporary literature on culturally related science and environmental education. Amidst this seemingly endless labyrinth of terms, it is easy to lose your way; a wide range of scholars provide an equally wide range of perspectives on societal trends of cultural interaction on regional, national, and international levels along with their beliefs of how these trends relate to current educational practices.

Multi, Inter, or Transcultural?

Welsch (1999) proposes that, in our increasingly interconnected world, many people's experiences are "transcultural" blends of several cultures simultaneously. Similar to Saul (2008) and Roth (2008), Welsch (1999), a German scholar, challenges what he describes as the archaic classical European belief in a one-nation, one-culture worldview. He suggests that past and present theories of multi- and interculturalism are also inadequate portrayals of the current complexity of regions, nations, and international relationships because they are based on the monocultural assumption of *difference* rather than recognizing the historical and contemporary ambiguity and blurring of cultural boundaries in most areas of the world.

Welsch (1999) proposes that a monocultural understanding of the world relies on three inherently flawed premises: social homogenization, ethnic consolidation, and intercultural delimitation. He states:

> Firstly, every culture is supposed to mould the whole life of the people concerned and of its individuals, making every act and every object an unmistakable instance of precisely *this* culture...*Secondly*, culture is always to be the "culture of the folk", representing..."the flower" of a folk's existence...Thirdly, a decided *delimitation* towards the outside ensues: Every culture is, as the culture of one folk, to be distinguished and to remain separated from other folks' cultures. The concept is separatory. (pp. 1–2)

Welsch suggests that these premises are inadequate because:

> First...modern societies are multicultural [internally culturally plural] in themselves, encompassing a multitude of varying ways of life and lifestyles...Secondly...ethnic consolidation [in the past] is dubious...such folk-bound definitions are highly imaginary and fictional; they must laboriously be brought to prevail against historical evidence of intermingling. Finally, the concept demands outer delimitation...The traditional [European] concept of culture is a concept of inner homogenization and outer separation at the same time...It tends...to be a sort of cultural racism. (p. 2)

While Welsch's perspective is unmistakably European, he does touch on several concepts that bear great relevance to this inquiry. Welsch emphasizes the internal plurality within all cultures of the past and present due to individual differences as well as interaction with other cultures. For example, in our increasingly connected world, we can almost instantly access food, clothing, literature, music, and languages from across the globe. Welsch also emphasizes the alienating effect of monoculturalism that defines itself, largely, by what it is not, requiring foreign "others" to consolidate one's own sense of culture.

Welsch (1999) also critiques the concepts of interculturalism and multiculturalism. He suggests that interculturality is inherently flawed because it relies on a monocultural view of the world in order to find a common middle ground. Welsch states:

> The concept of interculturality reacts to the fact that a conception of cultures as [independent] spheres necessarily leads to intercultural conflicts...the deficiency in this conception originates in that it drags along with it unchanged the premise of the traditional conception of culture. (p. 3)

Welsch (1999) also presents similar concerns about multiculturalism. He notes that the only distinction is multiculturalism's focus on intra-national rather than international diversity. Welsch proposes that:

> Compared to traditional calls for cultural homogeneity the concept is progressive, but its all too traditional understanding of cultures threatens to engender regressive tendencies which by appealing to a particularistic cultural identity lead to ghettoization or cultural fundamentalism. (p. 3)

Welsch (1999) suggests that transculturality is a much more appropriate tool to describe and understand cultures today. He argues that transculturality is a result of the "inner differentiation and complexity of modern cultures" which recognizes that "cultures today are extremely connected and entangled with each other" (p. 4). He also notes that, in many places, "cultures' external networking" (p. 4) has surpassed traditional internal beliefs and relationships. Welsch also notes that transcultural similarities have been enhanced through the spread of global movements such as environmentalism and feminism which have surpassed local or even national cultural concerns and created powerful sub-cultures whose members may have a more common outlook and lifestyle with their counterparts in similar sub-cultures of other nations than with other members of their home nation.

Welsch (1999) also suggests that cultures today are characterized by hybridization due to the relative ease and availability of "foreign" products and concepts. He states:

> Henceforward there is no longer anything absolutely foreign. Everything is within reach. Authenticity has become folklore, it is owness simulated for others—to whom the indigene himself belongs. To be sure there is still a regional-cultural rhetoric, but it is largely simulatory and aesthetic. (p. 5)

While I do agree with Welsch that the world has become increasingly complex and interrelated, his comments above concern me to a certain extent. Perhaps I am simply clinging to monocultural romanticism, but I find myself resisting his assertion that regional cultural characteristics are merely superficial decorations on top of a transcultural global cake. However, perhaps Welsch is merely observing what he perceives as a trend around the world for people to look increasingly outside of their respective region or nation to connect with products, concepts, and people around the world that appeal more to them than their own traditions. As someone who spent a large part of my twenties traveling and living around the world, I can understand this compulsion, but I find it troubling.

Reflecting on the concept of bioregionalism which emphasizes living in response to your local sociocultural and ecological surroundings (Dodge, 1981; Thomashow, 1996), I can't help but question the implications of a transcultural perspective for the ecology and/or cultures of specific regions. In a world where you can instantly download and read your favourite Romanian author on a laptop from the comfort of a cabin in northern Ontario, can you still value and support the local farmer's market? I believe that this *is* possible because I live this way and many others do too, maintaining virtual connections to other areas of Canada and the rest of the world while also valuing and participating in the local community. However, such a lifestyle requires an acutely critical consciousness to avoid becoming over-connected to the virtual world and disconnected from the local landscape.

I also question whether the relative ease of contemporary global travel and communication allows us to too easily wander about, place-less and disconnected from any meaningful engagement with a specific cultural and geographical region for any significant amount of time. Or perhaps, like the voyageurs who once travelled the breadth of Canada's boreal forest in a single season, some people now have an expanded sense of place that transcends the traditional interpretation of home as being limited to a narrowly defined geographical area (Cuthbertson, Heine, & Whitson, 1997).

As Grimwood, Haberer and Legault (2014) also describe, I have observed and experienced this myself as an outdoor educator. One characteristic of the outdoor education sub-culture is geographic and economic nomadism; contemporary outdoor educators not only travel great distances on canoe, ski, or hiking trips, but also, due the inherent seasonality of the work, across North America or even further abroad to take up new employment. For example, as an outdoor educator today, you might find yourself working as a canoe guide in Ontario for the summer and then as a backcountry ski guide in Alberta for the winter. Others might stick to one activity, necessitating raft guides to move south to the United States during winter, for example, in order to chase an "endless summer". Due to this seasonal nomadism, many contemporary outdoor educators feel at home in a variety of geographic settings.

Welsch's (1999) perspective is also inherently European; he is not writing from the perspective of someone who has been colonized and seen their culture erode under the forces of colonialism. From a historical perspective, I would argue that some cultures of the past were indeed quite different from one another. For example, pre-contact, the cultures of Europe were much more similar to each other than they were to, for example, North American Indigenous cultures. There is no question, however, that after contact, European and Indigenous cultures have influenced one another to varying degrees across North America (Saul, 2008). Perhaps a term such as "intercultural" can still be employed to describe the interaction of cultures that, at least in the past, *were* distinctly different from one another, while "transcultural" is more useful in a contemporary context.

Cajete (2000) and Snow (1977/2005) also remind us that, while there are many similarities amongst Indigenous peoples in North America, there is also considerable diversity. Similar to the unique, but related cultures of Europe, Indigenous North American cultures share many commonalities in worldview, cultural norms, and other characteristics, but also differ greatly in other ways. While generalities can be drawn about North American Indigenous cultures, indicating the precise tribal and regional origins of the material presented whenever possible helps to avoid "pan-Indian" over-generalizations. Saul (2008) suggests that embracing and working within this diversity is one of the strong traditions that Canada has inherited from our Aboriginal predecessors.

Welsch (1999) addresses these kinds of concerns when he clarifies that transculturality is not intended to homogenize or globalize out of existence all cultures of the past. Instead he proposes that it provides the opportunity for "new diversity". He states that:

> A new type of diversity takes shape: the diversity of different cultures and life-forms, each arising from transcultural permeations...The transcultural webs are, in short, woven with different threads, and in different manner...Transcultural networks... have some things in common while differing in others, showing overlaps and distinctions at the same time...No longer complying with geographical or national stipulations, but following pure cultural interchange processes. (pp. 9–10)

While I appreciate Welsch's optimism here for the potential of new cultural phenomena to occur, liberated from the shackles of historical monoculturalism, I disagree with his assertion that we must no longer comply with geographical stipulations. From an ecological and bioregional perspective, I find this troubling. As contemporary human beings, we still require physical sustenance from somewhere and, I believe, a sense of being at home some*where*. Despite being able to travel around the world by airplane in a single day and communicate instantly with people in distant lands, we are not merely particles floating in space, only interacting with each other regardless of our ecological and geographical surroundings.

Nevertheless, there is an alluring promise of hybrid harmony in a transcultural approach. If we were to interpret transculturalism on a regional level, I think that it holds great promise—as bioregional scholars (e.g., Dodge, 1981, McGinnis, 1999) suggest, would it not be ideal to see people in different regions embracing a collective culture developed in response to their unique geographical landscapes? The ecological benefits of such an approach are undeniable; for example, if more of us were to support local agriculture and harvesting, less fossil fuels would be burned transporting food to our area. Arguably less tangible, but equally important, benefits would also arise in phenomena such as increased community participation, transcultural cooperation, and the development of a collective sense of place.

Thunder Bay, a region where I spent many years under the watchful eye of the Sleeping Giant Nanabijou, on the western shores of Kichigami, also known as Lake Superior, is an example of an area that already has a strong regional culture based on an eclectic mix of original cultures (Scandinavian, Italian, Scottish, French, Ukranian, Anishnaabe, and Métis) brought together by a rugged and isolated landscape. In Thunder Bay there are several farmers' markets that are well attended and many people still value things such as local food, picking berries, hunting, and fishing. The various founding cultures have intermarried and actively share their traditions and practices with each other, resulting in a unique regional métissage. This is not to say that racism and prejudice do not exist; the local news often carries stories on shocking incidents of racism, especially towards Aboriginal peoples. For example, an elementary school teacher's aide was recently suspended amidst public outrage after she impulsively cut an Aboriginal boy's hair in class without his permission, despite widespread awareness at the school that he had been growing it for ceremonial reasons (CBC, 2009). However, despite ongoing incidents of cultural ignorance and prejudice, I do believe that this region provides an interesting example of regional transculturalism. There is most certainly a unique "feel" to the sociocultural landscape that extends south across the arbitrary United States-Canada border to northern Minnesota. As someone with eclectic ethnic roots who has travelled widely and lived in several places, I can relate to Welsch's (1999) supposition that:

> Transcultural identities comprehend a cosmopolitan side, but also a side of local affiliation...Transcultural people combine both. Of course, the local side can today still be determined by ethnic belonging or the community in which one grew up. But it doesn't have to be. People can make their own choice with respect to their affiliations. Their actual homeland can be far away from their original homeland. (p. 11)

Welsch (1999) also reminds us that:

> Conceptions of culture are not just descriptive concepts, but operative concepts. Our understanding of culture is an active factor in our cultural life. If one tells us (as the old concept of culture did) that culture is to be a homogeneity event, then we practice the required coercions and exclusions. We seek to satisfy the task we are set—and will be successful in so doing. Whereas, if one tells us or subsequent generations that culture ought to incorporate the foreign and do justice to transcultural components, then we will set about this task, and then corresponding feats of integration will belong to the real structure of our culture. The "reality" of culture is, in this sense, always a consequence too of our conceptions of culture. (p. 7)

Welsch's comments above are especially important for us to consider from an educational standpoint; as educators we have the privilege and power to influence our students through consciously or unconsciously relating our own our own worldviews. As Welsch asserts:

> One must therefore be aware of the responsibility which one takes on in propagandizing concepts of this type...Propagandizing the old concept of culture and its subsequent forms has today become irresponsible; better chances are found on the side of the concept of transculturality. (p. 7)

Welsch's (1999) concept of transculturalism relates strongly to métissage, the focus of the following section.

From Bricolage to Métissage

Similar to Welsch (1999), German-Canadian scholar and educator Roth (2008) also challenges the classical European notion of pure monocultures. His work explores the related concepts of *bricolage* and métissage and examines the experiences and identities of contemporary science educators and students. Roth comments:

> Cultural bricolage, taking from here and there to make do, producing not only new, heterogeneous, creolized forms of knowledgeability and practice but also producing hybrid identities in a process of continuous métissage. This métissage occurs in all parts of a society that—such as Canada—values multiculturalism and leads to new cultural phenomena. (p. 894)

Similar to Welsch's (1999) assertion that most cultures around the world are already *transcultural*, Roth proposes that due to the ever-increasing *diaspora* of many cultural groups, many people already live in a "métis" way, drawing from their home cultures and traditions, as well as those of their new homes. He notes:

> Ongoing globalization leads to an increasing scattering of cultural groups into other cultural groups where they…continue to be affiliated with one another thereby forming diasporic identities. Diasporic identities emerge from a process of cultural bricolage that leads to cultural métissage and therefore hybridity and heterogeneity. (p. 891)

Roth (2008) also makes the interesting suggestion that students raised in their "own" culture may still experience diaspora when they are placed in new educational settings such as science or mathematics classrooms that require them to view the world through foreign lenses. He states:

> *Diaspora* is a concept, therefore, that allows us to theorize not only the experiences deriving from transnational migration and how these mediate science learning but also the experiences of native students in a culture foreign to the one they experience at home. (p. 902)

Roth provides the example of a devoutly religious student confronted by a scientific worldview who feels confused and torn between the conflicting teachings of the student's family and church and those of the government-sponsored science teacher. Roth also explores the experiences of junior high French immersion science students. He notes that these students are in a double-diasporic situation—confronted not only with the new "language" of science, but also having to communicate among themselves and with their teacher in their second, or even third, language. Roth's observations helped me to understand some of my own experiences as a French immersion student; I can clearly recall challenging, but ultimately enriching, moments when I was required to learn new concepts in math or science classes. Conversely, I remember feeling out of place when attending English-language math and science classes later in high school and university because all of my foundational vocabulary was French. However, while these situations were challenging at the time, I believe that they were ultimately enriching because I am now capable of understanding complex concepts in either language. These experiences early in life also gave me the confidence to study other languages, such as Japanese, Ojibwe, and Chinook Wawa in subsequent years.

Roth (2008) comments that we can all experience these kinds of diasporic moments whenever we are placed in uncomfortable or unfamiliar situations and advocates for the learning and increased awareness that result from such experiences. He suggests that it is exactly this kind of diasporic thinking that is required to solve today's complex issues and challenges. He notes:

> Real-world problems (those outside a test tube) are so far ranging and complex that they require the métissage of resources from ranges of domains heretofore isolated within disciplines...we need individuals who can live a diasporic life and identity, employing the resources at hand to find solutions to problems that exceed the grasp of any single regional ontology. (p. 914)

Pieterse (2001) also proposes that métissage problematizes cultural boundaries—forcing people to re-examine their conceptions of culture and cultural identity. Similar to Welsch (1999), he comments that this kind of expanded understanding of cultural identity can also lead to a broader ecological identity. Pieterse (1996) also suggests that hybridity disrupts and confuses racist notions by blurring cultural and racial boundaries. Similar to Zembylas and Avraamidou (2008), Pieterse and Welsch (1999) relate that, despite negative connotations of racial hybridity in Western history, in more recent times hybridity has increasingly been viewed as positive. Pieterse describes hybridity as a form of post-modern subversion that can embody "a resurrection of subjugated knowledges" (p. 1393). However, he cautions against superficial hybridity, or "multiculturalism lite", and uses the metaphor of the mixing of words versus grammar in linguistic métissage: word hybridity would be a simple mixing of words to spice up a language, whereas grammatical hybridity would involve a deeply seated blending of the actual *structure* of a language.

Pieterse (2001) proposes that hybridity involves recognizing the "in-betweens" and "interstices" (p. 238) and pushes us beyond false dualistic conceptions of culture and race. According to Pieterse, this requires "collective liminality, collective awareness" (p. 239) similar to the Trickster knowledge celebrated in many Indigenous cultures.

Roth (2008) recognizes the "symbolic violence" experienced by minority and Indigenous students in classrooms around the world. He comments that science education is especially guilty in this regard:

> The violence does not come from the fact that they are different—difference exists inside White middle-class culture—but from the fact that enactments of their (inherently hybridized) cultural forms are repressed, punished, thereby leading to the (tacit, acknowledged) experience of oppression. More so, within White middle-class culture, legitimized, legitimating, and legitimate scientific discourses in particular—as scientists and science educators define them in their curriculum and Standards documents—are constituted and considered as superior to any hybrid discourse. Although science educators often give lip service to acknowledging the (cultural) pre-school and out-of-school experiences of students, they do so on the presupposition that it is a lesser form of knowing than the one to be inculcated. (p. 913)

Roth's comments on the tensions between Western science and Indigenous epistemologies lead us to the following section that examines current issues and practices in Indigenous science and environmental education.

Traditional Ecological Knowledge: Indigenous Science?

As has been discussed previously, a growing number of scholars have recently advocated for the development of the Third Space between Western and Indigenous ecological approaches (e.g., Barnhardt & Kawagley, 2005; Brandt, 2008; Roth, 2008; Zembylas & Avraamidou, 2008). Brandt (2008) comments:

> I am convinced that these locations exist within campus classrooms, in faculty offices, or in the corners of campus. *Locations of possibility* are those discursive spaces in which students and their instructors value connected knowing, acknowledge each other's history, culture, and knowledge. (p. 718)

These kinds of statements are inspiring, but what does enacting them in practice entail? Also, is Indigenous knowledge "science"? What are the similarities and differences between Western and Indigenous ecological knowledge and philosophies? Also, what issues do we need to be aware of before attempting to blend or integrate Western and Indigenous knowledge? The following section explores these kinds of questions.

Anishnaabe scholar and activist LaDuke (2002) emphasizes the ancient roots of TEK and argues for its inclusion in contemporary ecological discourse:

> Traditional ecological knowledge is the culturally and spiritually based way in which indigenous people relate to their ecosystems. This knowledge is founded on spiritual-cultural instructions from "time immemorial" and on generations of careful observation within an ecosystem of continuous residence. I believe that this knowledge represents the clearest empirically based system for resource management and ecosystem protection in North America…Frankly, these native societies have existed as the only example of sustainable living in North America for more than 300 years. (p. 78)

However, other Indigenous scholars express concern about how TEK might be included in mainstream dialogue and action. Describing the somewhat tense relationship between Western science and TEK, Anishnaabe scholar Simpson (2004) also states:

> Over the past fifteen years Traditional Ecological Knowledge (TEK) has received much attention in the United Nations–sanctioned forums concerned with biodiversity and sustainable development, and this has sparked the curiosity of scientists working in these areas. Those aspects of TEK that are most similar to data generated by the scientific method are seen as a potential resource, holding answers to the environmental problems afflicting modern colonizing societies, while the spiritual foundations of IK and the Indigenous values and worldviews that support it are of less interest often because they exist in opposition to the worldview and values of the dominating societies. (pp. 373–374)

Simpson notes that this interest in TEK was initially met with enthusiasm in some Indigenous circles:

> Initially, many Indigenous people viewed this new interest with optimism and hope, seeing an opportunity to indigenize environmental thinking and policy to the betterment of both Indigenous and non-Indigenous Peoples and to advance the agenda of decolonization and liberation. (p. 374)

However, Simpson suggests that many efforts to integrate TEK and Western science have failed due to the perpetuation of colonial attitudes by well-meaning Western scientists. She suggests that:

> This has not gone unnoticed by Indigenous Peoples, and interactions around TEK and resource management, conservation, sustainable development, and biodiversity have become important sites of resistance and mobilization for Indigenous Knowledge holders and political leaders advocating for Indigenous control over Indigenous territories and Indigenous Knowledge and promoting a decolonized and just approach to the coexistence of Indigenous and non-Indigenous nations. (p. 374)

Simpson also relates her frustration with what she describes as a continued lack of acknowledgement by Western scientists of the validity of TEK. She relates that much of her own research has been discredited as non-scientific and barred from publication in Western science journals. Simpson notes that mainstream interest in TEK stops when sociocultural questions are raised relating to, for example, *why* it is currently endangered (colonialism). In concert with other scholars (e.g., Reid, Teamey, & Dillon, 2002), Simpson suggests that Western scientists are reluctant to acknowledge the sociological, spiritual and cultural aspects of TEK. She (2004) argues that:

> Removing Indigenous Knowledge from a political sphere only reinforces the denial of the holocaust of the Americas and trains a generation of scientists to see contemporary Indigenous Peoples and Indigenous Knowledge as separate from our colonial past, as an untapped contemporary resource for their own exploitation and use. This serves as a reminder that it is not enough to recover certain aspects of Indigenous Knowledge systems that are palatable to the players in the colonial project. We must be strategic about how we recover and where we focus our efforts in order to ensure that the foundations of the system are protected and the inherently Indigenous processes for the continuation of Indigenous Knowledge are maintained. (p. 376)

Other Indigenous scholars also argue for the recognition of TEK as a valid form of science. Eminent Blackfoot scholar Little Bear (2000a, p. xi) states:

> If science is a search for reality and if science is a search for knowledge at the leading edges of the humanly knowable, then there are "sciences" other than the Western science of measurement. One of those other sciences is Native American science… In order to appreciate and "come to know" in the Native American science way, one has to understand the culture/worldview/paradigm of Native American people. (p. x)

Euro-Canadian scholar Snively (2009) supports Little Bear's view. She suggests that various forms of "science" exist in all cultures, reminds us that the original Latin root for science (*scientia*) simply means "knowledge" (p. 33), and states that "Indigenous science is an interpretation of how the world works" (p. 34).

Attempting to allay the concerns of those who question the validity of "Indigenous science", Kincheloe and Steinberg (2008) state:

> In this context, the Western analyst confronts the need to reassess the criteria for judging knowledge claims in light of the problems inherent in calling upon a transcultural, universal faculty of reason. Questioning and even rejecting absolute and transcendent Western reason does not mean that we are mired forever in a hell of relativism. (p. 137)

They describe the promise of what they term "transformative science":

> Once individuals come to believe that Western science is not the only legitimate knowledge producer, then maybe a conversation can be opened about how different forms of research and knowledge production take issues of locality, cultural values, and social justice seriously...The goal of such a learning process is to produce a transformative science, an approach to knowledge production that synthesizes ways of knowing expressed by the metonymies of hand, brain, and heart...A transformative scientist understands that any science is a social construction, produced in a particular culture in a specific historical era. (p. 153)

I believe that the most important aspect of this conversation is the recognition of TEK as a valid way of knowing and understanding the world without forcing it to conform to the norms and values of Western science. Snively (2009) and Little Bear (2000) argue that TEK is its own form of "science"; as Snively suggests, it is useful to distinguish *Indigenous* science from *Western* science as they most certainly descend from different cultural and methodological origins, but the root meaning of "science" is simply knowledge of how the world works. While Western science has come to denote a prescribed empirical process to approaching problems (e.g., hypothesis, testing, results, conclusions, further testing...), Indigenous science has its roots in a wider understanding of the world that includes disciplined observation of Nature, for example, but that is also enhanced through deeper spiritual and philosophical elements that extend to ontologies of daily life (Barnhardt & Kawagley, 2005; Simpson, 2004).

Stephens (2000) further compares and contrasts Western and Indigenous approaches in search of common ground. He suggests that there are many similarities between Western science and Indigenous knowledge of Nature including concepts such as a unified Universe, personal qualities such as perseverance, curiosity, and honesty, empirical observation of nature, and a desire to understand the behaviour and patterns of plants, animals and other beings and elements.

Differences that Stephens (2000) highlights include Indigenous trust in inherited wisdom contrasted with Western skepticism, Indigenous holism compared to Western compartmentalism, the Indigenous belief in the link between the metaphysical and physical worlds as opposed to the Western science focus on the physical world only, and the Indigenous tradition of seeking understanding in order to apply it to daily living versus the Western science value of seeking understanding for its own sake.

The Integrative Science program at Cape Breton University employs the Mi'kmaq concept of *Toqwa'tu'kl kjijitaqnn* or "Two-Eyed Seeing" to combine Western science with traditional Mi'kmaq knowledge and philosophies of Nature (Hatcher et al., 2009). The Integrative Science program's aim is "concentrating on common ground and respecting differences" to teach "both systems side by side" (p. 3). Bringing together Aboriginal and non-Aboriginal Elders, educators, and students, the Integrative Science program provides an interesting example of an integrated approach at the post-secondary level.

The leaders of the Integrative Science program (Hatcher et al. 2009) suggest that at certain times (e.g., measuring fish populations) Western science and mathematical approaches are best used, while at other times (e.g., teaching students trout behaviour, feeding habits, and how to catch them) an experiential Mi'kmaq approach is most suitable. Hatcher et al. suggest that, "In weaving back and forth between knowledges, Two-Eyed Seeing avoids a clash or "domination and assimilation" of knowledges" (p. 5). They also propose that the Integrative Science or Two-Eyed Seeing approach provides "fertile ground" (p. 4) for interdisciplinary educational projects. This approach is extremely promising and has proven successful for both Indigenous and non-Indigenous participants. With the current increase in Indigenous and non-Indigenous educators and students re-engaging with Indigenous knowledge, ethical issues have also arisen.

Engaging with Indigenous Knowledge

Conflicts over the collection and use of Indigenous knowledge have caused considerable debate over the past century. Academic fields such as anthropology and pharmacology have come into conflict with Indigenous peoples over intellectual property rights and the misrepresentation of cultures. As a result, Indigenous individuals and communities are often hesitant to share any knowledge with outsiders for fear of its misuse for profit or misrepresentation. As Cajete (1999) relates:

> It is important to move beyond the idealization and patronization of Indigenous knowledge, which often leads to marginalization of the most profound Indigenous ways of knowing...Indigenous people have been touted as the spiritual leaders of the environmental movement. Such a designation is more symbolic than tangible...Still, many environmental educators...[are] paralleling the traditional practices of Indigenous societies. This is appropriate since Indigenous people around the world have much to share and much to give. The same peoples also continue to be among the most exploited and oppressed, and are usually the people who suffer the greatest loss of self and culture when dealing with various economic...and educational schemes. (p. 19)

So, how can non-Indigenous people respectfully learn from and with Indigenous peoples? Root (2010) suggests that an important point to understand is that there are four different forms of Indigenous knowledge: traditional and sacred, revealed (through ceremony, dreams, and visions), empirical, and contemporary. Of these four, the kinds that *might* be shared publicly include empirical and contemporary. The other forms of knowledge are often restricted to family or community members who have earned the right to that knowledge. This is an important characteristic for non-Indigenous people to understand.

Another key issue introduced by Mack et al. (2012) is the fundamental importance of authentically integrating Western and Indigenous knowledge in educational programs. They suggest that:

> By integrating multiple ways of knowing into science classrooms, students will learn the value of traditional ways of knowing and Native language, learn to utilize a conceptual ecological perspective, and acknowledge that learning and understanding is part of a complex system that includes student experience, culture, and context, as well as mainstream materials that are taught in the classroom. (pp. 4–5)

While these kinds of statements seem initially promising, leading Indigenous scholars such as Battiste (1998; 2005), Simpson (2002; 2004), and Hermes (2000) contest the integration of Indigenous knowledge *into* previously established Western-style educational programs or curricula. Battiste, Simpson, and Hermes all suggest that, in order to avoid the subjugation or "watering down" of Indigenous knowledge, Indigenous educational programs must be developed from an Indigenous perspective *first* before considering how they might meet be tailored to meet Western standards, rather than the opposite.

Kincheloe and Steinberg (2008) acknowledge that engaging with these kinds of issues is a challenging task for educators:

> In this critical multilogical context, the purpose of indigenous education and the production of indigenous knowledge does not involve "saving" indigenous people but helping construct conditions that allow for indigenous self-sufficiency while learning from the vast storehouse of indigenous knowledges that provide compelling insights into all domains of human endeavour…Teachers and scholars informed by this critical multilogicality understand these concepts. Such educators and researchers work to extend their students' cognitive abilities, as they create situations where students come to view the world and disciplinary knowledge from as many frames of reference as possible. (pp. 135–139)

As Cajete (2000) concludes, in order to conduct respectful research and education, "an equal playing field is essential for exchange of information between practitioners of Indigenous and Western science" (p. 8).

Ecological Métissage

Considering recent comparisons of Western science and Indigenous knowledge (e.g., Kawagley and Barnhardt, 2005), if we expand our conception of "Western" environmental knowledge and philosophy to include approaches such as deep ecology (Naess & Rothenberg, 1990) and bioregionalism (Dodge, 1981), the distinctions between Western and Indigenous approaches diminish and more similarities emerge, positioning us well to move from bricolage to métissage.

Drawing from the traditions of deep ecology and bioregionalism allows us to include tenets such as respect and recognition of cultural and ecological diversity, the inherent value of all beings, spiritual forces, long-term multi-generational thinking, the embedded and relational position of human beings in the circle of life, locally focused and responsive living, practical application of theoretical principles, local traditions, and acknowledging Indigenous territories and sacred landmarks. Roth (2008) describes this shift from bricolage to métissage as a process, in this instance enacted as a move from intercultural integration to regionally inspired transcultural blending.

Figure 3 (below) is a graphic representation of ecological métissage. Similar to Figure 1, it was inspired by the infinity symbol found in the centre of the Métis flag to represent the joining of two cultures (Dorion & Prefontaine,

1999). Donald (2009) notes that, as with Métis peoples and our communities, every instance of theoretical, curricular or methodological métissage is distinctly developed through the unique lens of the individual theorist. The principles of ecological métissage are explained in further detail below.

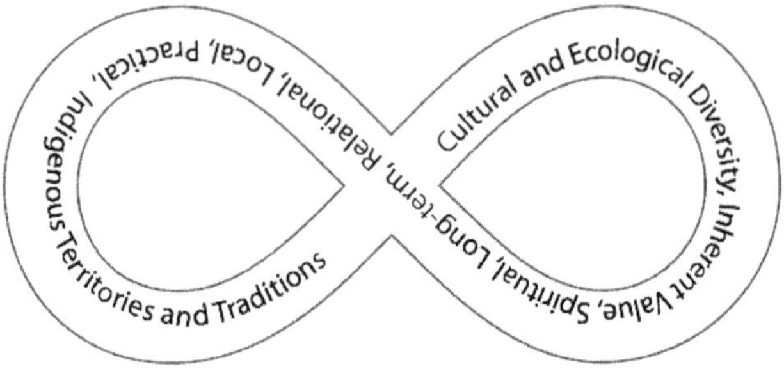

Figure 3. Ecological Métissage.

Acknowledging Indigenous Peoples and Territories

The central concept of ecological métissage is acknowledging and honouring Indigenous territories, as connection to the Land is a foundational characteristic of Indigenous cultures (Cajete, 1994; Myers, 2009). Central to this concept is recognition of the ancient symbiotic evolution of Indigenous languages, cultures, and worldviews within specific geographical areas. Several bioregional scholars also emphasize recognizing and engaging with Indigenous peoples, traditions and territories (Aberley, 1999; Dodge, 1981; McGinnis, 1999). While this perspective may have been somewhat shallow in the early years of bioregional scholarship, it seems to have taken on an increasingly authentic character.

The result of this recognition of and connection to specific geographical regions is similar to what many contemporary Western scholars call a sense of place, a strong feeling of being at home in a particular area (Curthoys, 2007). However, an Indigenous sense of place goes beyond simply feeling at home or connected to the Land. As Snow (1977/2005) explains from a Nakoda Sioux perspective:

> The Rocky Mountains are sacred to us...These mountains are our temples, our sanctuaries, and our resting places. They are a place of hope, a place of vision, a place of refuge, a very special and holy place where the Great Spirit speaks with us. *Therefore, these mountains are our sacred places*. (p. 19)

This sacred sense of place, relation, and interconnection is explained further through the concept of "All My Relations".

All My Relations

Inherent in most Indigenous worldviews is recognition of the inherent value, spirit, and interconnectedness of all people, living creatures and bioregions (Cajete, 1999). The cyclical and interconnected worldview of Indigenous North American cultures is often illuminated in the concept of "All My Relations" (Simpson, 2004). As Cajete notes, "an ecological sense of relationship encompassed every aspect of traditional American Indian life" (p. 4). According to Cajete, this respect for and deep sense of relationship with the greater-than-human world results in a feeling of kinship and responsibility as "Indigenous people [feel] responsibility not only for themselves, but also for the entire world around them" (p. 11).

Similarly, deep ecology also argues for a more humble humanity; it exhorts us to experience the world as a whole, recognizing the inherent rights of other people and beings and living accordingly (Naess & Rothenberg, 1990). As Drengson (2008) explains, "deep ecology supporters appreciate the inherent value of all beings and of diversity" (pp. 8–9). The late founder of deep ecology, Naess, encouraged us to view ourselves as *part of* the web of life on a physical as well as a moral and spiritual level (Naess & Rothenberg, 1990). In a manner similar to many Indigenous traditions, deep ecologists advocate for a more embodied, respectful, and reflective approach to ecology, placing humans in the circle of life with the rest of the natural world, rather than outside of it as detached "shallow ecologists" (Stibbe, 2004, p. 243).

Respect for Cultural and Ecological Diversity

Recognizing that all of Creation is inherently valuable also means honouring cultural and ecological diversity. As deep ecologist Macy (2007) explains:

> Diversity is a source of resilience. This is good news because this time of great challenge demands more commitment, endurance, and courage than any one of us can dredge up out of our own individual supply. (p. 29)

Drengson (2008) also notes that:

> The...principles [of deep ecology] are supported by people from diverse backgrounds...They each have their own personal ecosophy...This sense of global solidarity helps us persist in our efforts. Exactly what policies and actions we undertake depends on our personal situation, cultural context, and individual place. *No single solution can be applied to every place*. (p. 30)

This emphasis on the strength of regionally grounded diversity is reminiscent of Indigenous scholars such as Snow (1977/2005) who comments, "The creator created diversity amongst plants, animals, and people. So isn't diversity a good thing?" (p. 23).

Locally Grounded, Long-Term Vision

Locally grounded, long-term vision and planning is another characteristic common to Indigenous, deep ecological, and bioregional traditions. For example, deep ecologist Drengson (2008) suggests that "abstract knowledge is not sufficient for a full life" (p. 40). Aberley (1999) also emphasizes a locally relevant, practical approach when he notes that:

> Bioregionalism is best understood when viewed from the "inside," not from reading one or several texts. Gatherings should be attended, ephemeral periodicals reviewed, restoration projects participated in, and place-based rituals and ceremonies shared. (p. 32)

In a similar spirit, Anishnaabe scholar and activist LaDuke (2002) states, "you take only what you need, and you leave the rest" (p. 80) while describing the Haida tradition of harvesting lengths of timber from living trees without chopping them down as an example. This kind of long-term thinking embodied in action is characteristic of Indigenous ecological wisdom and is often represented by the Seven Generations maxim (Coates, Gray, & Hetherington, 2006): How will our actions today impact our descendants seven generations from now?

Examples of Contemporary Programs

Adams, Luitel, Alfonso, and Taylor (2008) note that:

> There comes a time in the development of theoretical perspectives...when one is entitled to ask for a demonstration of their practical viability in assisting resolution of problems, dilemmas and challenges confronting teachers and students. (p. 1002)

The same authors also suggest that "one of the most important roles of educational theory is to help radically reconceptualize" (p. 1002) the taken-for-granted assumptions of modern science and environmental education.

Given the previous discussions of the search for a deeper understanding of concepts such as métissage, transculturalism, diaspora, and the Third Space mentality, what are their practical applications in contemporary science and environmental education? This is an emerging concept that has been explored and employed recently by a growing number of scholars and educators. Variably naming their programs "Integrative Science" (Hatcher & Bartlett, 2009a), "Cross-Cultural Science" (Snively, 2009), "Two-Eyed Seeing" (Hatcher, Bartlett, Marshall, & Marshall, 2009), or "Bridging the Gap" (Kazina & Swayze, 2009), the collective aim of these practitioners is to embody the Third Space in educational practice. Third Space approaches are often interdisciplinary, providing opportunities to integrate science, social studies, humanities, art, health, and physical education into one unit or lesson. My own experiences as an outdoor, environmental, and science educator in both formal and informal learning environments resonate with this approach. In the following I begin with a discussion of programs designed specifically for Indigenous students and end with exploration of those designed for students from all backgrounds. International examples are also provided.

Indigenous Environmental Education Programs

Simpson's (2002) description of her experiences as an Indigenous environmental educator in Canada is one of the most comprehensive descriptions of Indigenous environmental education available. Based on her extensive experience with various programs, Simpson relates several key features that she believes should be part of any Indigenous environmental education program: supporting decolonization, grounding programs in Indigenous philosophies of education, allowing space for the discussion and comparison of Indigenous and Western epistemologies, emphasizing Indigenous ways of teaching and learning, creating opportunities to connect with the Land, employing Indigenous instructors as role models, involving Elders as experts, and using Indigenous languages when possible. Simpson's recommendations provided the framework for my master's study (Lowan, 2007, 2008, 2009) that explored Outward Bound Canada's Giwaykiwin program for Aboriginal youth through a lens of decolonization.

Another study that I first encountered in my master's work was Takano's (2005) description of a community-developed land-based cultural education program based in Igloolik, Nunavut. Takano, a researcher of Japanese descent, participated in Paariaqtuqtut, a 400 km journey through the community's ancestral territory in May 2002. Paariaqtuqtut means "meeting on the trail" in Inuktitut and was developed by a group of community members and Elders. Paariaqtuqtut aims to connect young people with cultural skills and teachings in a land-based context.

Takano (2005) found that the community members in Igloolik were concerned that many youth were losing connections with their land and culture. Those interviewed observed that this leads to youth feeling lost between two worlds, disconnected from their community and culture, yet unprepared to live in the Western world. Takano also recorded the experiences of several participants who felt that Paariaqutuqtut had helped them to reconnect with their land and culture.

Barnhardt and Kawagley (2005) also describe the Alaska Rural Systemic Initiative (AKRSI), a program developed in 1995 in consultation with Indigenous Alaskan Elders. The intent of the program was to enhance Indigenous students' experiences in the formal school system by integrating local Indigenous values and knowledge into school curricula across Alaska. Barnhardt and Kawagley report significantly improved educational experiences and academic success for participating students. While the AKRSI was originally designed for Indigenous students, it has since been employed to include students from all cultural backgrounds across Alaska. As Barnhardt and Kawagley note:

> By documenting the integrity of locally situated cultural knowledge and skills and critiquing the learning processes by which such knowledge is transmitted, acquired, and utilized, Alaska Native and other Indigenous people engage in a form of self-determination that will not only benefit themselves, but will also open opportunities to better understand learning in all its manifestations, thereby informing educational practices for the benefit of all. (p. 20)

The AKRSI now serves as a catalytic model for other programs across North America. The following section explores other such programs that reach out to not only Indigenous students, but those from other cultural backgrounds as well.

Intercultural Environmental Education Programs

Hatcher and Bartlett (2009a) of Cape Breton University's Integrative Science program describe a series of lessons that they often employ on local birds for Indigenous and non-Indigenous students alike. Using a Two-Eyed Seeing approach, they introduce students to the various birds of their area through a combination of experiential and classroom-based activities drawing from various fields of Western science along with local Mi'kmaq legends and knowledge. Their lessons include everything from observations of birds at bird feeders, discussions of the physics of flight, local Mi'kmaq beliefs and knowledge about different birds, and exercises on storytelling and the oral tradition. They suggest that their "classroom mirrors the world outside" (p. 7) and also that "Integrative Science is underlain by…an intimate, respectful relationship between the scientist, the natural world and different worldviews" (p. 9).

Hatcher and Bartlett (2009b) also provide a clear example of a Two-Eyed Seeing lesson that teaches students about local medicinal plants using a combination of Mi'kmaq ethnobotanical knowledge of the healing properties of tamarack and red spruce and chemical testing for the presence of Vitamin C. The lesson culminates with the students preparing a drinkable tea.

Métis educators Kazina and Swayze (2009; see also Swayze, 2009) describe their program, "Bridging the Gap", that serves inner-city youth in Winnipeg. Bridging the Gap aims to impart a sense of ecological stewardship to its elementary-aged students through a combination of Western science, experiential, and Aboriginal approaches. Kazina and Swayze describe their two main lesson approaches: The first involves an introductory talk by an Elder; students and teachers alike are introduced to important cultural protocols such as how to properly approach an Elder: providing a small offering of tobacco beforehand as a traditional request for the sharing of knowledge. Kazina and Swayze suggest that involving Elders is a comfortable way for non-Aboriginal educators to respectfully introduce Aboriginal concepts to their students. Following the introductory Elder's talk, students are taken on an interpretive hike where they get hands-on experience and learn traditional protocols for activities such as picking medicinal plants and berries. The lesson culminates with the students recording their observations in a journal.

Another lesson begins again with an introductory discussion with an Elder on the Aboriginal concept of "All My Relations", the interconnectedness of all things (Durst, 2004; Pepper & White, 1996; Simpson, 2004).

Students then work in groups to create "habitat wheels" that illustrate the various needs of animals and how they fit into their respective ecosystems. The Medicine Wheel, a common symbol used by many Aboriginal peoples to represent universal interconnection (Cajete, 1994, 1999; Snow, 1977/2005), is used as the base framework for the habitat wheel. This cyclical, interconnected model contrasts with the common hierarchical Western concept of the food chain or pyramid that I remember from my early school days.

Trent University in central Ontario has also gained acclaim for its Indigenous Environmental Studies program (Evering & Longboat, 2013). Trent's approach involves a combination of strategies aimed to revive and support Indigenous traditions and beliefs in program design, delivery, evaluation and application outside of the university.

Snively (2009), a professor at the University of Victoria on Canada's Pacific coast, describes her own "cross-cultural science" approach. She states:

> Cross-cultural science education is a topic that either polarizes or numbs people depending on their understanding of the concept and their agenda for science education. Those who think there is only one right answer and one definition of science may think that cross-cultural science education is fundamentally flawed and a waste of time. Those who tend to believe that we can approach questions from different angles and starting points and still come up with workable solutions usually think that cross-cultural science is imperative. (p. 33)

Snively works with students from university to grade school level. Her approach to creating a Third Space for students includes challenging them to solve a problem (e.g., constructing a fishing net) using only their creativity and local resources. Once they have completed their task, she reveals the local Indigenous technology that was historically developed in response to the same problem and uses this as an entry point for discussion around the meaning of "science". Snively closes by saying:

> Cross-cultural science education is not merely throwing in an Aboriginal story, putting together a diorama of Aboriginal fishing methods, or even acknowledging the contributions Aboriginal peoples have made to medicine. Most importantly, cross-cultural science education is not anti-Western science. Its purpose is not to silence voices, but to give voice to cultures not usually heard and to recognize and celebrate all ideas and contributions. (p. 38)

There are also inspiring examples of international Indigenous outdoor and environmental education programs. For example, Lertzman (2002) and

Henley (1989) describe the Rediscovery program, a global family of outdoor and environmental education programs based on local Indigenous traditions. Henley (1989), one of the program's founders, states, "Rediscovery brings together people from many different racial backgrounds...when people from different races have the opportunity to talk to one another, to work and play together, then inevitably they begin to learn about each other's lives and cultures" (p. 35). Rediscovery programs have been founded across North America and around the world (e.g., Wales, Thailand, Bolivia, Guyana, Siberia, Hong Kong) in various forms. Some are very small and focus on one particular Aboriginal community while others such as Ghost River Rediscovery (Lertzman, 2002) in Canada are large, year-round programs.

Hawaiian educator N. Wilson (2008) also discusses leading traditional Polynesian *waka ama* (canoe) voyages with Maori and non-Maori New Zealanders. Wilson reports that the inter-Indigenous context of these voyages facilitates rich intercultural ecological, metaphysical, and epistemological discussions and experiences for students and leaders alike.

Oguri (2010) of Kagoshima University in southern Japan also reports on a "living village" developed in Minamata, a small city famous for its remarkable recovery from widespread mercury poisoning in the 1950's, designed to revive, preserve, and share traditional farming, fishing, and forest skills, knowledge, and beliefs. Oguri reports that the citizens of Minamata have been surprised by the interest shown by people from larger urban areas who now regularly visit their "living village" to learn traditional skills and philosophies that have been lost in other areas of Japan.

One key experience that I had was at an Aboriginal high school in northwestern Ontario. I had the opportunity to spend six weeks working with students from grades nine to twelve in an integrated traditional land skills program. The philosophy of the school was to simultaneously impart traditional Ojibwe and Oji-Cree language, knowledge, and philosophies to the students while enhancing their Western science, math, social studies, and language skills. The hope of the teachers was that students would feel an enhanced sense of pride in their own culture as well as feel prepared to venture to southern Ontario and elsewhere for post-secondary schooling.

Our classes were often spent entirely outdoors in the surrounding boreal forest under the expert guidance of two local Elders. Students learned traditional land skills such as how to build shelters using local materials, make fire from various sources, set animal traps and fishing nets, and many other valuable lessons. Throughout the process students were encouraged to film and photograph their projects for final written and web-based reports (for examples go to: http://www.nnec.on.ca/pffnhs/traditional-activities).

One lesson that I was involved with developing and delivering serves as a great exemplar of the integrative approach of this program: First, the students selected a plant from the surrounding area and documented it by harvesting (after offering tobacco), drawing, or photography. Their next challenge was to learn the names for the plant in their Indigenous language (Ojibwe or Oji-Cree), Latin, and English. Students then researched the possible uses of the plant. Legitimate research sources included formal Western-style books and resources as well as local knowledge-holders or Elders. This approach emphasizes the equal value of both ways of knowing and gathering information. It fully legitimizes Aboriginal knowledge alongside Western sources. This lesson also provided a great opportunity to discuss current issues such as the misuse and appropriation of Indigenous plant knowledge for profit by non-Indigenous corporations.

The programs described in the previous passages embody a Third Space approach to science and/or environmental education. A Third Space approach challenges traditional Western academic boundaries, pushing the limits of "science" while embracing a transcultural or Métis mentality. Snively (2009) suggests that this means more than simply including the odd Aboriginal fact or example, it requires a collective movement into a Third Space where all cultural forms of "science" or knowledge are considered equally and implemented as is appropriate in the local context. Roth (2008) highlights the need for further research into the facilitation of Third Space thinking in contemporary science and environmental education. Zembylas and Avraamidou (2008) also recommend deeper inquiry into making science and environmental education more socially conscious and locally responsive.

While there is a growing body of literature on intercultural environmental education in Canada and internationally, no comprehensive studies to date have focused on the experiences and competencies of intercultural environmental educators working in Indigenous contexts and the deeper societal implications of their work. Who are these "border crossers"? What led them to their chosen vocation? What makes them effective? And how might they be reshaping Canadian ecological identity?

. 5 .

ENVIRONMENTAL EDUCATORS' PERSPECTIVES

I am very grateful to have had the incredible opportunity to engage in conversation with some of Canada's most experienced and inspiring intercultural environmental educators. As mentioned in earlier chapters, this journey was a constant process of reflection—a spiraling dialogue between the participants, the literature, and myself. As I came to the end, I found that some of my original questions had been answered, however new questions had inevitably risen in their place. Another conundrum that I faced was that every participant had a unique perspective on the topics and questions guiding this study. How then, to make sense of it all?

I realized that I needed to revisit the literature, discuss my thoughts with trusted peers and mentors, and reflect on my original research questions to begin making sense of all of these ideas and questions that had arisen during the interviews. My original research questions were:

1. Can Western and Indigenous knowledge of the natural world be blended theoretically and in practice? If so, how?
2. What characterizes the ecological identities of contemporary intercultural environmental educators?
3. Do they embody ecological métissage? If so, how?

4. How might the concept of ecological métissage [and/or similar concepts] reshape environmental education in Canada?

I also revisited the methodologies that guided me in the initial stages of this research such as Kovach's (2010) thoughts on Indigenous methodologies and the various interpretive and narrative researchers who informed my approach. As previously discussed, my intention in this study was also to seek out and explore participants' "epiphanic" (Denzin, 1989) or "aha" type moments that help us to gain deeper insights into the experiences, and perspectives of participants and ourselves.

The Participants

Before discussing my own summary insights, it seems most appropriate to provide further background on the ten educators who generously agreed to participate in my doctoral study. In my original dissertation (Lowan, 2011b), each participant's fully interpreted narrative was presented individually intact; this required an additional number of pages equal to the entire length of this book. So, like Kovach (2010), I hope that the ancestors will forgive me for the abridged versions provided below.

Skywalker[1]

Skywalker is a middle-aged woman from southern Alberta of British and Norwegian descent who runs an outdoor and environmental education consulting company in the eastern foothills of the Canadian Rockies. She has close to thirty years of experience facilitating experiential education with a wide range of students and has cultivated relationships and partnerships with Aboriginal individuals and communities throughout her career. Skywalker and I met on a lovely July afternoon; I bicycled to our meeting and we conducted our interview over fresh berries next to a slow-moving river. Our discussion was regularly interrupted by the singing of birds.

Natasha Little

Natasha Little is an Acadian woman in her early thirties of French and Mi'kmaq ancestry. She has been involved in the field of intercultural environmental education for over a decade. Natasha first became involved with intercultural environmental education on Canada's East Coast in a university-level integrative science program. She studied and worked there for almost a decade before graduate studies took her to British Columbia. Natasha has since returned to the East Coast and now works with a non-profit environmental education organization. We met on a hot July afternoon at her home and conducted our interview in her backyard. Our only distractions were birds and the delicious aroma of salmon roasting on the barbecue.

Orange Blossom

Orange Blossom is a middle-aged woman of Swiss, Dutch, and German descent originally from the American Pacific Northwest. She relocated to Canada's West Coast Early in her teaching career where she has been an intercultural science and environmental educator with a wide range of Indigenous and non-Indigenous students for over forty years. We met at a café in a suburban community on the West Coast. It was a fresh, early summer day, so we decided to sit outside, despite steady background noise from vehicle traffic nearby.

Liatrus

Liatrus is a woman from a mixed cultural background (Métis, Pakistani, Euro-Canadian) in her early thirties with over a decade of experience as a student and practitioner of environmental education. Amongst many other commitments, she is currently the director of an urban environmental education program in a large prairie city that teaches both Western science and Indigenous perspectives. Liatrus and I met on a muggy summer evening inside a busy café. We had a wonderful conversation that flowed very naturally amidst the buzz of other patrons and café staff. As will become evident in the following, Liatrus raised several interesting points of discussion that kept me thinking long after our interview had concluded.

Olivia L'Hirondelle

Olivia L'Hirondelle is a Métis woman in her late twenties. She is currently the program coordinator of an urban environmental education program for Aboriginal youth that incorporates both Western science and Indigenous perspectives. Olivia and I met early on a late August morning at her office in a large prairie city. The conversation began at a slow and sleepy pace, but became increasingly animated as our morning coffees took hold.

Thor

Thor is a man in his mid-thirties of mixed European descent originally from southern Ontario. He has been involved in outdoor and environmental education with a wide variety of programs across Canada and internationally for over ten years. Over the past few years, Thor has worked extensively with Inuit youth both in formal classroom settings as well as on extended wilderness canoe trips. We held our interview after dark on a fall evening next to a crackling outdoor fire. The smell of fresh rain and wet leaves filled the night air. Thor provided many thoughtful, unique, and challenging responses that fuelled our discussion and introduced new perspectives to this study.

Kate

Kate is a woman in her early thirties of mixed European descent. She has been involved with outdoor and environmental education as a participant, leader, and teacher educator for over fifteen years. The current focus of her teaching and research is reflexively examining the experiences of White outdoor and environmental educators who are engaging with Indigenous peoples and knowledge through a lens of decolonization. Kate and I held our interview over Skype, an online video conferencing program, after several failed attempts to meet in person. We have known each other for several years so, despite being physically separated, our conversation flowed smoothly. We both commented afterwards that it was easy to forget that we were not speaking in person.

Tong

Tong is a western Canadian woman in her mid-twenties of mixed European ancestry. Her involvement with intercultural outdoor and environmental education began over ten years ago as a youth participant with an intercultural outdoor and environmental organization that welcomes students from all cultural backgrounds. This initial experience motivated Tong to return several times and gradually led to her to taking on a leadership role with the same organization. She continued her immersion in outdoor and environmental education as an undergraduate student. Tong currently works as a school librarian in a northern community. Similar to Kate, Tong and I conducted our interview using video chat technology. We've also known each other for several years, so our conversation flowed quite naturally despite the physical separation.

Arthur

Arthur is a man in his early thirties of Japanese and Danish descent who grew up in a large urban area in central Canada. Arthur's interest in outdoor and environmental education began in his late teens when he enrolled in an undergraduate outdoor recreation program and has culminated with him running a successful dogsled and bed and breakfast business in northwestern Ontario for the past four years. Arthur draws on a soft-spoken yet highly engaging teaching style to facilitate experiential learning that incorporates local Aboriginal knowledge, perspectives, and partnerships. We held our interview in a cozy living room on a chilly afternoon late in the fall.

Cedar Basket

Cedar Basket is a Sto:lo educator and young Elder from Canada's West Coast. She has been an educator in a wide variety of settings for over twenty years and currently teaches various courses in Aboriginal and technological education at a university in central Canada. We met in a quiet office on a snowy mid-winter's day.

In the following I highlight epiphanic ideas and discussions that emerged during conversations with this incredibly rich group of educators.

Exploring Intercultural Environmental Educators' Ecological Identities

As with any group of people identified as having a certain commonality such as, in this case, a profession, the contemporary intercultural environmental educators interviewed for this study represented a wide range of backgrounds, experiences, opinions, ideas, and identities. However, they also exhibited similarities such as the common perspective of seeing themselves as *part of* rather than *separate from* Nature, as well as significant life experiences (SLE) such as regular and extended time spent outdoors as children or young adults, influential teachers or family members, and academic study.

For example, Skywalker, a Euro-Canadian woman, stated, "I…perceive myself [as] part of Nature…" and described regularly spending time in Nature with her family as a child as highly influential on the development of her adult ecological identity. Liatrus, a woman of mixed cultural heritage (Pakistani, Aboriginal, and European), also commented, "I see myself as being a part of [Nature]" and described time spent camping with her parents as a child along with academic study later in life as highly influential on the development of her current ecological identity. Similar stories emerged amongst all of the participants. Some, such as Orange Blossom, acknowledged a familiarity with significant life experience research when she commented:

> I…look back on [my childhood experiences] and…I know from having read…lots of environmental education articles [that] usually there is a teacher or a role model. I had neither of those.

Instead of discussing an influential adult in her early years, Orange Blossom reflected on her free-range childhood, where she was given space to explore the forests and coastlines of the Pacific Northwest, as the primary influence on her early love for Nature. I will not dwell on findings such as these that have been explored in detail in past significant life experience research (e.g., Dillon, Kelsey, & Duque-Aristizábal, 1999; Palmer, Suggate, Robottom, & Hart, 1999; Payne, 1999; Thomashow, 1996). However, vastly different perspectives and experiences emerged when I considered how or if the educators in this study embodied ecological métissage.

At the outset of this study I defined ecological métissage as the blending of two or more ecological worldviews at a personal and/or cultural level as represented in personal identity, philosophies, and practices. This definition provides the framework for the following discussion that explores the identities, philosophies, and practices of the intercultural environmental educators in this study. As will be explored in this section, I argue that while some of the participants did embody ecological métissage in a variety of ways (identity, philosophy, practice), others expressed characteristics more congruent with bricolage—embodying an integrated rather than a blended perspective. The relationship between participants' cultural and ecological identities was also a highly variable area of discovery; while some expressed strong cultural affiliations that influence their ecological identities, others did not.

Ecological Identity

As noted, all of the educators that participated in this study identified themselves as "part of Nature". However, as each discussion progressed, it became clear that the influences on and expressions of their identities were quite diverse. Intriguing dialogue emerged on topics such as the influence of physical and geographical contexts on ecological identity as well as the relationship between cultural and ecological identity.

Context

Dillon et al. (1999) emphasize the dynamism of identity; they observe that a person's identity can shift from one *social* context to the next. An interesting and related tension raised by both Tong, a Euro-Canadian woman, and Liatrus, a woman of mixed Pakistani and Métis descent, is how shifting physical or geographical (e.g., urban or non-urban) contexts can influence their *ecological* identities; both Tong and Liatrus expressed feeling less connected to Nature in urban environments. For example, Liatrus commented:

> Living in the city...You start to have this sense of domination of Nature where you're very...urban and removed from...the natural setting. But I've started to look at appreciating any sort of human construction as a part of Nature, just in a very sort of humanized state, right... It all came from something and it all originated from something... That being said...I can't even describe the sort of emotional sensation to me to just be outdoors and be around sort of the typical Nature setting...by a river, in a forest...It's almost a healing; It de-stresses; It's just sort of that sense of being... back home...

I appreciate Liatrus' and Tong's honesty in sharing their struggles to maintain a sense of connection while in urban environments. I can definitely relate to their feelings; however, I agree with Liatrus that since humans are part of Nature, so too are our constructions. This concept is reminiscent of ecosopher Snyder (2003) who also argues that all human creations are, in fact, part of Nature, but agrees that time spent in areas less impacted by humans can be sublime. Indigenous scholars such as Cajete (1994) also remind us that all land (including contemporary urban areas) is a sacred part of Nature. What other connections might there be between culture and ecological identity?

Culture and Ecological Identity

Some participants in this study identified a strong connection between their cultural and ecological identities, while others emphasized the influence of geographical regions and subcultures such as professional peer groups (e.g., outdoor educators). Tong, a Euro-Canadian woman, also shared that her experiences with Aboriginal cultures have led her to rediscover the Earth-based traditions of her own European cultures. Skywalker seemed to concur when she commented on the influence of her Celtic and Scandinavian roots:

> I think that I've always felt a deep connection to the Celtic side of being [British]... that Celtic sort of...mystery...has always been a strong interest...Whereas...in terms of the Norwegian part... I love the cold and I love the heat...the sauna and the snow...

However, similar to Thor and Kate, both Euro-Canadians, Skywalker identified with being "Canadian"; she also emphasized her love for and connection to the Canadian landscape when she noted:

> I see myself as a Canadian...I've always felt...the [pride] of being Canadian and for me that [pride] has always been about the landscape, the wilderness, the wildlife, you know, the fact that all of that still exists in Canada, is part of what makes me a proud Canadian...And if that goes I'll be a disappointed Canadian!

Similar to several other participants (i.e., Skywalker, Thor, Arthur), Natasha Little also highlighted the regional variations of Canadian cultural and ecological identity. Based on her experiences growing up in an Acadian family on Cape Breton Island, she identified the inextricable relationship between the Acadian landscape and culture:

> I [don't]...draw lines between my ecological...and my cultural upbringing...I see... a party with my family as an ecological lesson and I see...spending time outdoors or learning about the environment as a cultural lesson as well. So I really don't... separate them. I think [that] because I grew up spending so much time outdoors...we [were] never "learning about ecosystems"...because it was often with my family...it's kind of always tied together with the cultural aspect...

From a West Coast Sto:lo perspective, Cedar Basket, an Elder and academic, also identified the powerful link between her family, community, and their home river when she commented:

> Well, our community is Sto:lo...The River and the people have the same name... And...my family...lived right beside the River so, culturally I identified with that River...I felt very lucky to [have] grow[n] up in such a rich way...

Many of the non-Aboriginal participants in my doctoral study seemed very cautious about appropriating an Aboriginal identity, while others shared stories of being welcomed by Elders into Aboriginal cultures. For instance, Kate, a Euro-Canadian, shared a perspective that involves deepening her understanding of the effects of colonization and Indigenous knowledge and philosophies, and then applying them to her own life without appropriating them into her *identity*. She commented:

> One thing I have to learn is…how to work towards resisting ongoing colonization and Eurocentrism…learning how to recognize the…patterns in my own thinking and in the thinking of people from my culture…And then the constructive side too of… just trying to begin to glimpse another way of viewing or understanding the world and that's not to say that I think we're out to necessarily completely take on an Indigenous worldview or…to learn all the knowledge that there is to learn…just… learning to recognize that there…are some other ways of relating to…other people and to Nature and to our history…

Arthur, a Canadian man of mixed Japanese and Danish ancestry, also described feeling welcomed into an Aboriginal culture by an Elder who shared a special story with him at a northern storytelling festival when he reflected:

> I realized that she was actually sharing something with me, that her telling me a story wasn't about her speaking, it was about her giving something to me…Now I was a part of that culture and I held that story…and…that…made me a part of it, it invited me in. And…it changed my thinking about culture in general.

Natasha, an Acadian of French and Mi'kmaq ancestry also related the story of a Mi'kmaq Elder telling her nephew that being Mi'kmaq was a way of seeing the world, not just an ethnic identity:

> [Elder] said, "Oh [nephew]! Being Mi'kmaq has nothing to do with what's in your blood, it's what's in your heart!"…It was a great lesson for both of us… It doesn't matter what percentage of your blood [is Aboriginal]… could be zero percent of your blood, but if you identify, if you choose to identify in a certain way then…who will stop you from doing that? If you're righteous about it then that's different, but if you choose to take on a worldview it's not the same as staking claim over something…it's choosing to look at something from a certain point of view…

One Métis participant, Olivia, eloquently expressed her own ecologically "Métis" identity:

> I mean they're both [European and Aboriginal] *me*...It's all a part of me...I think of myself as a relative of all things, animate and inanimate...My knowing of the Earth is that it is a whole in which all living is included. Creator and Mother Earth must be respected...It is a type of insanity to see Nature or the environment as "out there".

Kate, Liatrus, and Arthur all problematized the concept of "culture". Along with discussing the legacy of their ethnic identities, they also identified the influence of subcultures and regional cultures on their ecological identities. For example, Liatrus noted that "culture" can be interpreted differently by different individuals. She also emphasized the complex and evolving nature of her own identity as a person of mixed ancestry:

> "Culture" can have many different contexts. To some it means ethnicity...or race. And to some it means something completely different...It's actually more challenging for me because my cultural and ethnic history are so complicated...My cultural identity is kind of an evolving thing...

Kate also questioned the meaning of "culture" and described feeling very connected to the outdoor educator sub-culture:

> I think of...outdoor education culture...and how those are probably some of the... traditions that I cherish a lot in terms of...rhythms of the year like meeting up with people at a certain time...

Reflecting a bioregional perspective, Arthur commented on his strong and interconnected sense of connection to the distinct culture and geography of northwestern Ontario, his adopted home:

> I find that...my culture's very much...shaped by northwestern Ontario's landscape... It means I gravitate towards water...I embrace the snow...I like things slower and... the people I lean towards as role models are people...who have done things a little bit more traditionally...I'm the type of person who likes to be out on my ash snowshoes instead of my high-tech ones and...the dogs I use for dog sledding are older style lines...So culturally I would say, I've got a lot of influences...But...the way I found myself is more through how I interacted with the landscape...I suppose that's the best way to...describe my culture...

Exploring the relationship between the cultural and ecological identities of the intercultural environmental educators in my doctoral study revealed a variety of influences, expressions, and opinions. While some, such as Cedar Basket and Olivia, identified strongly with their cultural heritage and expressed how it is intimately linked to their ecological identity, others, such as Kate and Arthur, also highlighted the influence of other factors such as sub-cultures of like-minded people and geographical regions.

The question of whether or not it is appropriate or even possible, for someone to adopt a cultural identity different than their own raised a plurality of opinions. While some, such as Kate described her belief in *learning with* but not adopting Indigenous cultures, others such as Natasha and Arthur described experiences with Aboriginal Elders that led them to believe that one can be "welcomed" into another culture if they handle themselves with deep respect and humility, challenging the boundary between philosophy and personal identity. I am inclined to optimistically align myself with the perspectives of Natasha and Arthur; while I am extremely cognizant and respectful of the hesitation of many Indigenous peoples to open up to sharing culture and knowledge with non-Indigenous people (e.g., Simpson, 2004), I am hopeful that, through a continued increase in authentic and mutually respectful engagement, more people such as Arthur will be "welcomed into" Indigenous communities, further expanding opportunities for intercultural and eventually transcultural (Welsch, 1999) experiences and creations grounded in a regional sense of collective connection.

Philosophies

The ten educators involved in this study expressed a variety of eco-pedagogical philosophies; some expressed more of a Third Space perspective than others; however, they all acknowledged both Aboriginal and Western influences on their pedagogical values and practices. As was mentioned in the previous section, while many of the non-Aboriginal participants were cautious about adopting Aboriginal or Métis *identities*, they seemed more comfortable engaging with and adopting Indigenous philosophies to varying degrees. For example, similar to Natasha's comments about her nephew and their Mi'kmaq Elder, Liatrus commented:

> There's a lot of non-Aboriginal people that have a much better understanding of Aboriginal traditions than Aboriginal people. So it's not always something that's connected to our...ethnicity and race.

Skywalker, a Euro-Canadian woman, demonstrated such an understanding when she mentioned the importance of seeking the guidance of an Elder before visiting a sacred valley with students when she stated, "I knew that I wanted to have an Elder know what I was doing and...get their blessing...

So...I did go out [and I] met a fellow...[I] told him about my ideas and my interests...For me going to...this man...was really important."

Orange Blossom also discussed her experiences with incorporating both Western and Aboriginal influences into her personal and teaching philosophies. She mentioned that, along with having a background in the sciences, her pedagogical praxis is informed by and open to the arts, philosophy, and spirituality, allowing her to act as an intercultural and interdisciplinary "border crosser" (Hones, 1999):

> You know, I see things differently because I don't have that science degree...even though I did take a number of biology and oceanography courses. Really that's one thing that has allowed me to kind of squeak in the middle and kind of cross over...I was very much into values of worldviews, ethics, philosophy, art, and a wide range of interests...And that's why the Indigenous viewpoint appeals to me because [of] values...[and] ethics [and] spirituality... So I do see things differently...

Cedar Basket also described two epiphanic moments from her work with Aboriginal student teachers when she shared the ultimate realization they came to as a group that it *was* possible for them to bring Indigenous perspectives into their teaching in a predominantly Western system:

> [In the end] they [felt that,] "Yes, we're in this Western, still very powerful Western-looking curriculum and yet, no, there's a place at the table for what we're bringing forward here."...And that was just fabulous! And so, as an educator, I see how important it is...to really create an environment where they...have confidence about the worldviews they bring, about those stories from the Land...

Cedar Basket also described her approach to working with non-Aboriginal student teachers that involves encouraging them to consider Aboriginal knowledge and history alongside Western narratives and worldviews in a manner similar to Donald's (2009; 2010) conception of Indigenous Métissage (see Chapter Four):

> The very first thing I [do]...is to say that, "There has...always been...another narrative, an original narrative...here in this land, across Canada. Before...contact, before the Hudson Bay [Company], there was a narrative that was deep and rich and it's still here today. It hasn't gone away...And I believe that...when new teachers are welcomed in...where they don't feel like it's their fault what happened in the past with Canadian history, that they have a very powerful place to play now...And for the most part I find that they're leaving...with a...lot of resources and maybe somewhere down the line they'll still be thinking about some of the conversation that went on in that course.

Cedar Basket's description of how philosophy can inform practice leads well into the following section that examines the practices of the educators who participated in this study. How are their cultural and ecological identities expressed in practice? Do they embody ecological métissage?

Practices

How are intercultural environmental educators' cultural and ecological identities and philosophies related to their educational practices? The educators who participated in this study discussed these relationships both explicitly and implicitly. For example, Liatrus clearly described the relationship between her own mixed cultural background and her current practice as an intercultural environmental educator when she stated:

> My current research and professional interests are so closely intertwined with my... personal identity...And it's definitely influenced a lot of the choices that I've made in life, just being open to this idea of intercultural...ideas and that kind of thing... and being sensitive to...others' opinions and so on...

Liatrus also shared an epiphanic moment from her master's work where she realized that she needed the guidance of an Aboriginal Elder to support her Western-style academic research. Natasha's description of her teaching approach in an integrated science program also reflected a combination of Western and Aboriginal approaches:

> Well, for [many] years I was working with [a university-based] integrative science program...So I was kind of looking out there to see what was happening in Indigenous education...and then also keeping an eye on what was happening...in the science world and trying to bring them together. So [for example] there would be a new article out on dreams...and in the class we would be working on consciousness, so we would talk about...what's happening physiologically in the brain when you're dreaming and then also have a conversation or do an activity on visions or dreams culturally...

Orange Blossom also reflected on her approach of opening science education students' minds to recognize TEK as an equal of Western science when she stated:

> On the first day of class I'd get them to write their definition of science down on a piece of paper, put it in an envelope and put their name on it, and...And then I give

> them those envelopes back at the end of the course…It's amazing how their definitions change, sometimes significantly, at the end of the course.

Liatrus, Natasha, and Orange Blossom's comments remind me of Barnhardt and Kawagley's (2005) concepts presented earlier, comparing and contrasting Western science and Traditional Ecological Knowledge. Barnhardt and Kawagley suggest that, even though there are differences between Western and Indigenous approaches, there are also many similarities. They propose that understanding these distinctions is the key to drawing on the strengths of both. The following section explores the perspectives of other study participants in response to the overarching question: Is it possible to blend Western and Indigenous knowledge of Nature? If so, how?

To Blend or Not to Blend: From Bricolage to Métissage?

The overarching question framing my doctoral study was: Can Western and Indigenous knowledge of the natural world be blended theoretically and in practice? If so, how? As the study progressed, a variety of perspectives emerged on the concept of blending Western and Indigenous knowledge and philosophies.

Echoes from the literature resonated throughout our discussions; while some participants were strong proponents of *transcultural* métissage—seeking new blended, hybrid creations that draw on the strengths of both to create something new, others recommended a more cautious *intercultural* approach reminiscent of bricolage and Kimmerer's (2013) assertions; they suggested that, while there are many points of convergence, Western and Indigenous knowledge can and should never be completely blended. They suggested that the similarities between Aboriginal and Western knowledge and philosophies can be highlighted and explored, while still recognizing and honouring their distinctions. The different opinions presented by the participants interacted with the perspectives of various scholars and my own thoughts and opinions to create a dynamic dialogue.

Reflecting the concerns of Anishnaabe scholars Hermes (2000) and Simpson (2004) who caution against the "watering down" of Indigenous knowledge, Olivia and Liatrus expressed reservations about the absolute blending of Western and Indigenous knowledge in a predominantly Western context.

Olivia expressed her support for an integrated, but not blended, approach when she noted:

> In bringing Western and Aboriginal philosophies and knowledge of Nature together, I don't think that blending is the correct way to focus. I think that they can learn, share and teach from each other and there are ways in which they are coming together, but to blend is to blur and for me this is something that we need to be aware of...To honour it, you need to honour where it comes from.

Liatrus also challenged the use of terms such as "blend" that imply mixing from the beginning rather than seeking commonalities:

> I don't think you can use the term "blend". I think you can use them both together... [but] you're definitely not blending...I've always had very strong opinions of that...I mean my work specifically...involves Indigenous...and [Western] science education... and then sort of bringing in the environmental education piece into that. But I kind of came to a place where I decided...I wanted to spell it out...using [both] Indigenous knowledge and pedagogies and science...or ecological education, but [with] each remaining distinct.

Expressing a similar hesitancy to Olivia and Liatrus, Kate questioned the implications of the terms "intercultural" and "transcultural" when she stated:

> I also think you have to ask what you mean by "blending"...I'm thinking about the [difference] between inter-disciplinarity and trans-disciplinarity..."Inter" being... working in the spaces between and looking for connections between two disciplines. And "trans" being bringing two together to create something new...So...are you talking about meshing two cultures into one new culture or are we talking about looking for ways for two cultures to co-exist in certain spaces or certain people?

Similar to Indigenous scholars and educators such as Battiste (1998, 2005) who caution against simply integrating Indigenous knowledge *into* Western-style curricula, Liatrus also emphasized the importance of equal representation of Western and Aboriginal content that begins in the planning stages of program development. However, similar to Roth's (2008) discussion of the spontaneous and creative nature of bricolage that eventually results in métissage, Liatrus also commented:

> I can almost think of a piece where you could do blended instruction if it wasn't preplanned...but just in sort of the day-in-day-out informal, like if [I] was just walking down the street, and talking to somebody and I'm...teaching them something it's

going be a blended, because it's going to be...all the different sort of ethnic...and cultural things that I know...

Natasha also discussed the spontaneity that she experienced while teaching in an integrative science program in the Maritimes. She raised the interesting point that blending can be very effective, but at other times, keeping Western and Indigenous approaches separate is better. Being sensitive to these nuances is a key characteristic of effective intercultural environmental education practice. Natasha commented:

> Sometimes we would do strictly science and sometimes we would do [a] traditional [Indigenous] activity...And sometimes they blended really well and sometimes...you didn't need to blend them...Sometimes it would water them down if you blended...and we didn't want to water down science or...be an assimilation...We wanted it to be an "integration" or coming together and...recognizing that there are differences that shouldn't be blended.

Kate expressed similar concerns to Olivia and Liatrus for the watering down of cultures that might occur through unabated blending, but her comments also express similar sentiments to Natasha's perspective on the nuances of when blending may or may not be appropriate:

> I don't think I could say...that everything should be blended because then...the only option is to create something new...As opposed to...looking at what kinds of traditions to conserve that...have worked for a really long time...But maybe where blending is useful is...as a bridge for people to find points of resonance and find places where they can get on the same page and form some relationships and have some conversations...I mean I don't think it's realistic to...completely abolish one way of thinking or another...[However]...figuring out what to do with these different ways of knowing is...important...

In a manner reminiscent of Barnhardt and Kawagley's (2005) previously discussed comparison of Western and Indigenous knowledge and philosophies of Nature, Natasha discussed the unique strengths that she perceived in each tradition—the holism of Indigenous approaches and the attention to minute details facilitated by Western science approaches:

> Well, I think that spirituality is at the base of...Indigenous...culture. There's a deep...spiritual connection. And...as it stands, Western science cannot fit spirituality into its realm. It doesn't...[acknowledge] the need for...spirituality. It...dismisses or...takes away from it...And same with the...strictly objective reductionist components of...Western science...Sometimes you don't need to see the big picture,

> you just want to look at the details…It's funny…over the past few years…I've been working in these two-way…programs that claim…to be working in a co-learning [environment]…And I think that reductionism…[is] often painted with a bad stroke…in a negative sense…But learning to look at details is really important too… Western medicine is what it is because we learned to look at small, small details… But I think it's lost without the big picture…I think it's pointless to look at details without the big picture…

Sto:lo scholar and educator Cedar Basket expressed her optimism for the future of blending and transculturalism when she stated:

> I really want to see [a] meshing of science and of art and of literature and of math… It's…not even interdisciplinary, it's *trans*disciplinary…It's going beyond just how, there they are, all linking, but what comes out of that?…If we're really serious about education and revitalization in the second decade of the 21st century…we need to start exploring…different ways of how we put knowledge and curriculum together. And yes, Aboriginal worldview[s] and science…are [are] all…part of that…

Skywalker also shared an interesting story about her experiences working with children who are recent immigrants to Canada, emphasizing her belief in being open to learning with and from her students and the rich complexity that emerges when Western, Indigenous, and other worldviews are brought together:

> I was doing a program at the zoo on…interaction in ecosystems, and we're looking at all these native ecosystems and I realized as I looked around…it wasn't a classroom of Albertans who grew up on or near farms or had a Grandma on a farm. They didn't know about gophers or badgers. Nobody had even heard of a badger…These kids knew more about tigers and…Eurasian animals…because that's where most of [their families] were from…So that was an interesting paradigm [shift] for me…Because I was teaching them about…Canadian animals and they were teaching me about… other continental animals…So then…how do I teach so that it's easy for them to build the bridge? I need to think out of the box…I need to be flexible enough to realize they bring these other worldviews and to somehow find a way to make that bridge so that they're still feeling like they know something, which they do, and they're [still] understanding the essential concept of what they need to know and they're not wrong because they don't know about cougars…

Cognizant of the concerns of participants such as Liatrus and Olivia and scholars such as Hermes (2000) and Simpson (2004) surrounding the watering down and misappropriation of Indigenous knowledge, I would suggest that in the case of environmental education initiatives that bring together Western and Indigenous knowledge and philosophies, the best approaches begin as an intercultural, integrated *bricolage* of two or more epistemological and ontological approaches, being careful to recognize and discuss the original sources of the knowledge that you are presenting. As Sto:lo scholar and educator Cedar Basket enthusiastically advocated in her interview, this process will hopefully result in an expansion of the Third Space and the proliferation of locally grounded transcultural instances of métissage, as Roth (2008) also suggests.

However, considering persisting concerns regarding the misinterpretation and misappropriation of Indigenous knowledge, I think that it is important that we continue to exercise caution to articulate, understand, and honour the distinctions and similarities between Western and Indigenous science. My position is based on voices in the literature as well as my experiences over several years of teaching introductory Aboriginal education courses for undergraduate students; the students whom I work with typically do not arrive with a comprehensive understanding of historical or contemporary issues facing Aboriginal peoples in Canada today, nor do they demonstrate a comprehensive understanding of Aboriginal cultures. This experience leads me to believe that we must still work towards deeper intercultural understanding before considering the potential for regionally grounded transculturalism. Perhaps, as Natasha and Liatrus suggested, the most promising approach involves learning to recognize those concepts and knowledges that best lend themselves to transcultural blending or mixing (métissage) and those that are best kept as cautiously integrated instances of intercultural bricolage.

I am not certain what the future holds, but my hope is that in future decades we will find ourselves in a different sociocultural situation in Canada where concerns regarding misappropriation have softened and a critically informed intercultural bioregionalism will have taken hold. Such a shift will also require a broader conception of environmental education to include and embrace Indigenous as well as Western perspectives other than science such as deep ecology and bioregionalism. Based on my own experiences as a Métis scholar and educator and those related by the participants in this study, I believe that this is possible and, in fact, increasingly becoming a reality. I am inspired by the continued growth of intercultural science and environmental education programs; as documented in this study, new programs and initiatives are constantly emerging across North America. Perhaps it is not too late for us to cautiously, but enthusiastically, revive the Third Space, the Métis spirit that was commonplace in the early days of our nation, but lost in recent generations.

As will be discussed in the following chapter, Cedar Basket and Skywalker's comments and encouragement led me to conduct further conversations with three more participants who had come to Canada in their youth. I was specifically interested in understanding their experiences related to learning about Western and Indigenous ecological knowledge.

Note

1. All participants selected their own pseudonyms.

. 6 .

THREE-EYED SEEING?

As previously discussed, Canada is a culturally complex country composed of Indigenous peoples and settler populations from Europe and, increasingly, other parts of the world. In general, but with some important exceptions, the first waves of Canadian colonizers and settlers were from Europe, bringing with them predominantly Western perspectives on science, ecology, and land use (Saul, 2008). For the first few centuries of post-contact Canadian history, these Western perspectives interacted and often clashed with Indigenous understandings of the natural world based on thousands of years of geographically rooted experience (Cajete, 1994). More recently, immigration from other parts of the world has increased (Malenfant, Lebel, & Martel, 2010). As Skywalker suggested, people arriving from non-European cultures might have an understanding of Western science and philosophy, but they also often carry rich ecological understandings linked to their home nations. Statistics Canada projects that immigration from non-European countries will continue at a high rate over the next several decades (Malenfant, Lebel, & Martel, 2010). Simultaneously, Indigenous history, perspectives, and contemporary issues are increasingly emphasized in many Canadian provinces and territories as priority areas in education for all students.

As discussed in the previous chapter, such trends have created and revealed rich and wonderful pedagogical complexity for Canadian educators and students alike. While there is extensive literature available pertaining to multicultural science and environmental education (e.g., Agyeman, 2003; Blanchet-Cohen & Reilly, 2013; Roth, 2008) and a growing body of work on Indigenous science and environmental education (e.g., Aikenhead & Michell, 2011; Cajete, 1994; Elliot, 2011; Hogue, 2012; Lowan-Trudeau, 2012a, 2013a; Snively & Corsiglia, 2000; Scully, 2012; Sutherland & Swayze, 2012; Swayze, 2009), research that examines the complex interaction of these two areas is limited. As previously discussed, concepts such as "Two" or "multiple-eyed seeing" (Bartlett, 2005; Hatcher, Bartlett, Marshall, & Marshall, 2009; Institute for Integrative Science and Health, 2012; McKeon, 2012) have been introduced to describe these complex contexts, but they have not yet been explored in great detail together in the Canadian context.

As such, motivated by conversations with Cedar Basket, Skywalker, and other participants in my doctoral research, I conducted a short follow-up pilot study to gain insight into the complex experiences of newcomers to Canada with learning about Indigenous ecological knowledge in predominantly Western educational contexts.

Towards Three-Eyed Seeing?

As outlined in Chapter Four, the concept of Two-Eyed Seeing is now well established in science and environmental education circles. Developed by a team of Mi'kmaq Elders and researchers at Cape Breton University, Two-Eyed Seeing involves viewing the world simultaneously through one Western and one Indigenous eye to form a balanced and unified whole (Bartlett, 2005; Hatcher, Bartlett, Marshall, & Marshall, 2009). This concept has proven very useful and is adaptable to a variety of cultural and geographical contexts engaging Western and Indigenous ecological knowledge.

The developers of Two-Eyed Seeing allude to the possibility of other culturally rooted perspectives being considered in addition to Western and Indigenous knowledge (Institute for Integrative Science and Health, 2012), however no empirical research to date has explored the potential of Three-, Four-, or Five-Eyed Seeing in earnest.

As such, the purpose of this pilot study was to explore the formal and informal educational experiences of first-generation immigrants to Canada regarding Indigenous ecological knowledge and philosophy. Specifically, I was guided by the following questions:

- How do newcomers to Canada perceive Indigenous ecological knowledge and philosophy?
- How might formal and informal science and environmental educators better respond to such culturally complex educational contexts?
- What are the broader societal implications of these kinds of questions?

In order to consider such questions, this pilot study also employed methodological métissage (Lowan-Trudeau, 2012b) as discussed in Chapter Two, a calculated mix of interpretive, narrative, and Indigenous research approaches. Three pilot interviews employing a semi-structured format were conducted with first-generation adults who were born elsewhere, but had experienced some kind of formal schooling in Canada. In the spirit of a pilot study, sample size was intentionally kept very small in order to allow for in-depth consideration, interpretation, and presentation of participants' narratives.

Participants and Recruitment

Despite broad circulation of a call for participants to appropriate community and professional networks, I did encounter initial difficulty with recruiting participants. During reflection and discussion with participants and colleagues, it was proposed that this may have been due to the relative paucity of adult individuals new to Canada who have had the opportunity to engage with Indigenous ecological knowledge in meaningful formal or informal educational contexts. In fact, this foreshadowed one of the key findings of this study explained in further detail below.

Another surprising methodological development was that, despite the recent increase of immigration from non-European countries (Malenfant, Lebel, & Martel, 2010), there were, of course, individuals from European nations who were interested in participating in this study. This was an important reminder of the ongoing complexity of immigration in Canada. As presented below, this resulted in a participant group that loosely mirrors historical Canadian immigration trends.

In order to provide further insight into the participants and their perspectives, brief biographies, in chronological interview order, are presented below:

Kathy[1]

Kathy was born in Oxford, England. Early in childhood, her family left England by ocean liner for Canada, eventually settling in Ottawa in 1967 where she still lives today. Kathy noted that, as a British immigrant, she found it fairly easy to transition into life and school in Canada. In her current position, she manages an Aboriginal youth mentorship program that brings Aboriginal role models into communities to facilitate sport, leadership, health, and development initiatives.

Sophia

Sophia is currently a post-doctoral researcher in the natural sciences. She was born in central Europe and, similar to Kathy, came to Canada by ocean liner in early childhood. Sophia's family initially settled in Ottawa, but soon relocated to a small lakeside community in central Ontario. Sophia shared that, overall, she had a comfortable childhood in a predominantly Anglo-Canadian community at the time. However, she did experience some prejudice and feelings of exclusion related to her family's central European cultural and linguistic roots.

Takwana

Takwana is an educator currently living in Thunder Bay, Ontario. She came with her family from Zimbabwe to Toronto in her mid-teens, a difficult time for such a transition. Despite experiencing significant prejudice and culture shock, Takwana successfully completed high school and undergraduate studies in southern Ontario. She was also employed for several summers as a literacy and community development worker in several Aboriginal communities across northern Ontario. This experience led Takwana to relocate to a university in northern Ontario to complete her graduate studies.

Interpretation

In a manner similar to my doctoral study, interviews were transcribed, re-storied (Creswell, 2002), and individually and collectively coded for themes (Lichtman, 2012). Each interview was also examined for epiphanic, illuminating, or "aha" moments where participants and/or the researcher experienced exceptional clarity or understanding. In the spirit of reciprocity common to Indigenous research methodologies (Kovach, 2010), in-depth and individually intact narrative portraits (Lawrence-Lightfoot, 2005) were subsequently produced and presented to each participant. In recognition of individual and community accountability, I maintain regular contact with the participants and seek their approval and insight regarding any publicly presented or published materials.

Key Insights

These three conversations produced an incredible depth and diversity of insights, experiences, and perspectives. However, as a pilot study, this is just the beginning of a much deeper line of inquiry. Notable findings from this study include the common lack of meaningful exposure to Indigenous knowledge and philosophy of any kind through formal schooling, the importance of critical and experiential approaches, impacts of these experiences on participants' identities, and the potential for re-imagining cultural complexity as a strength rather than deficit for collaboratively addressing contemporary socio-ecological issues through formal and informal education.

Limited Exposure to Indigenous Knowledge

All three participants emphatically stated that they had very little exposure to Indigenous ecological knowledge and philosophy in their formal K-12 education in Canada. However, Takwana reflected upon her earlier experiences with school in Zimbabwe where both the English and Shona (local Indigenous culture) languages and cultures were naturally integrated into school curricula and the day-to-day functioning of the school and community. For example, she shared memories reminiscent of Two-Eyed Seeing (Hatcher, Bartlett, Marshall, & Marshall, 2009) and traditional Métis approaches to education and religion discussed in Chapter Three when she noted:

> In elementary school [in Zimbabwe], we [had] a garden...and sometimes we'd do class projects where we'd be growing things...We learned English and Shona...We would read Shakespeare, and...Nigerian authors like Chinua Achebe...I also...remember...when we were learning animals, the class would be learning [about local] Shona [and European] animals...It wasn't that people were trying, that's just how life was...Having a grandmother come in and tell stories to the class first thing in the morning...that was just something that was done...If you were sick, she would get traditional medicines and then pray to [Christian] God that those traditional

> medicines would work…Whenever she was travelling, she would give me salt [and herbs] to put around the house to protect us…and she would say, "remember to keep the Bible under your pillow". So, it was just a mixture of things. It wasn't formal training like "this is traditional and this is not"…It was both.

However, Takwana also reflected on a common reticence amongst youth in her community that differed from her observations of Indigenous youth in Canada:

> Few young people wanted to admit to those things happening…everybody wanted to be listening to Usher and TLC and…nobody wanted to participate in anything traditional [or] learn the traditional instruments because it was like, "ew, that's old". I notice…when I go to some reserves here…there's never a rejection of like, "Oh Christianity did this to us…" It's just kind of an acknowledgement of…accepting that spirituality and [also] having…traditional ways of doing things…and…I appreciate when I see the kids here at pow-wows…I'm like, "yeah!" they're not ashamed of everything that's not Western like I was growing up.

Despite such reticence in her early years, due to being raised surrounded by both Western and Shona cultures and languages, Takwana alluded to the possibility of Three-Eyed Seeing when she expressed feeling more open to Indigenous traditions in Canada. She commented, "When I think about myself and when I came here as an Indigenous person from elsewhere, it might [have been] easier for me to accept Indigenous knowledge here." However, along with Kathy and Sophia, Takwana expressed frustration that she was not exposed to Indigenous peoples and knowledge during her early years in Canada. She talked about searching well into her undergraduate and early graduate studies for mentors and opportunities to express, explore, and relate her own Shona culture to Indigenous peoples in Canada.

Fostering Socio-Critical & Experiential Approaches

Takwana's emphasis on experiential, community-based pedagogies in Zimbabwe aligns well with Indigenous perspectives here in Canada as described in the literature (e.g. Elliot, 2011; Lowan, 2009, 2013b: Simpson, 2002) and by the educators who participated in my doctoral study. All three participants in the pilot study spoke of the formative influence of opportunities that they had to spend extended time in Indigenous communities in Canada for work or post-secondary study.

For example, both Takwana and Kathy discussed their overwhelmingly positive experiences working and spending time on the Land in several Indigenous communities for extended periods of time. Kathy, originally from England, commented:

> I'm a firm believer in experiential learning and...opportunities for students to get out on the Land...to get out and...see and do and touch and smell...and to share foods, take part in ceremonies...Giving them some context...so that they're better able to appreciate it more holistically.

Sophia, who emigrated from central Europe in early childhood, shared similar sentiments when she reflected upon her experiences as a student researcher in the Canadian North:

> I went up north for my first field season and I had a really, really profound time... I stayed on an island with two families [where] I was the only White person...and... it was my first chance to really...spend time with Indigenous people...It was really amazing to be...immersed in this different culture... My eyes were just wide open and I was quite quiet and really just observing everything.

Reflecting further on these experiences, Sophia also suggested that experiential, land-based approaches might prove successful in introducing new Canadians to Indigenous knowledge:

> We all as individuals have...a lot of knowledge about the area that we come from and sometimes, if you don't go outside very much, it will be a very small area...but we all know that area around us and we can always learn more in different ways...about the area around you, the Land, the people, the wildlife, the fish, the water, everything... and...that crosses all cultures and I think that's really exciting [and] something...we can all relate to.

When I asked Sophia to further consider how science and environmental educators might respond to culturally complex learning contexts that involve Indigenous knowledge, she initially underlined the historical lack of authentic Indigenous perspectives in Canadian education. She also emphasized the importance of such learning for all students and shared several insights linked to current societal dynamics based on her experiences:

> I think it's imperative that we have this for everyone. [There is still a lot of] racism within our country...It's really disturbing and...I think our current government, not [supporting] cross-cultural understanding and respect is totally political. It's...meant to oppress and...it's working...but I don't think it's going to work much longer and

> I feel really excited about Idle No More[2] and [other related events]...It's really an amazing time [and] I think it gives people the capacity to communicate and to... educate...and that's what I'd like to see.

Sophia's comments and my own experiences with the pedagogical opportunities inherent in Indigenous social and environmental activism encouraged me to develop a subsequent study that will be described further in later chapters.

All three of the participants in the pilot study agreed that it was crucially important for non-Indigenous formal and informal learners of all ages engaging with Indigenous knowledge and peoples to have strong mentors who are able to facilitate respectful and critical intercultural exchange and dialogue.

Identity Transformation

Similar to many of the educators in my doctoral study, all three participants in the pilot study also spoke about how their experiences in Indigenous communities transformed their identities. For example, Kathy shared that she now sometimes finds it hard to relate to her British relatives' perceptions of Indigenous peoples in Canada. When asked to share her thoughts on how she has come to understand Indigenous ecological knowledge and philosophy in Canada through her lens as an English person, Kathy noted distinctions between Indigenous and British approaches to the Land, plants, and animals. Similar to Skywalker, she lamented the loss of Indigenous knowledge in the United Kingdom. For example, she commented:

> The U.K. is so completely different...traditional medicines and teachings and you know, living off the land and that type of thing...it's just not something that is part of... British culture and traditions [anymore]...Yes they sort of go on a hunt every now and then, but this is an upper-class British tradition that involves a lot of dogs and such...as opposed to the traditions and teachings that go along with First Nations' communities that come together around a hunt...That...difference...is so stark...

Kathy also reflected upon her British relatives' impressions of her work with Indigenous peoples in Canada:

> A couple of years ago a number of my...relatives came over...and...I was talking about what it is that I do and who I'm working with and...I could see that it was just a completely foreign concept to them, you know, what does it mean to be a First Peoples and...live in this kind of environment...with all the resources and traditions... and...different things. It was so completely foreign to them. I don't think that they actually understood what...I was talking about...

Sophia's insights were reminiscent of the internal struggles described by several of the educators in my doctoral study; she described shifts in her identity as she moved from respecting, but not fully accepting Indigenous knowledge during her undergraduate studies, to fervently adopting Inuit perspectives after her first extended experiences in the North, to finding a point of balance where Western and Indigenous knowledge and philosophies comfortably co-exist. Sophia reflected:

> [I was] romanticizing and…trying to reject everything that I was and just be like [the Inuit]…It's funny looking back…I was quite young I guess and just…really mesmerized by this entirely different culture.

However, over time she came to a more balanced perspective:

> Learning to respect these different ways of knowing is really important and…quite powerful. I can't speak for [the Inuit people]…I can only try and understand what's been explained and what I've read…It can be really awkward because then I've interpreted what I've been told and…I'm trying to somehow…not play the devil's advocate, but…be sensitive to different ways of knowing. I get kind of lost in all of that and…that's part of…my identity…It makes me kind of a messy person! Trying to navigate…who I am and…where…I come from and what…I know…from a very scientific perspective, but then [I'm also] really informed by everything that I hear and learn…Every time I go up North…things make more sense.

Sophia's romantic notions of Inuit peoples and epistemologies and rejection of Western ways were further tempered as her experience in the Arctic and understandings of Inuit perspectives deepened; she began to critically compare and contrast her own personal assumptions and beliefs with both Indigenous and Western science and ecosystem management approaches. For example, referring to beluga whale and polar bear conservation, she noted:

> We had a lot of conversations where some of my views were realigned about…hunting and harvesting…Having…conceptions [such as]…is it fair that you harvest a female who's pregnant? I would say things like…"Oh, I feel really sad for that fetus that's never going to get born" and then…the Elder…would just say things like, "Well…it's really not a big deal…the population is healthy and we don't need to have this romantic view…it's just the way it is."

Similar to several of the participants in my doctoral study, Sophia also reflected on her increasing consciousness relating to her personal identity and understanding of Indigenous knowledge:

> I'll never be able to have an Indigenous way of knowing…I see things based on my own cultural heritage and my family…upbringing…[However] some people…feel they can really, even if they're non-Indigenous themselves, they can see things from an Indigenous perspective after enough time and…enough teachings. [However] I guess I don't believe that for myself…

Takwana also described an epiphanic or "a-ha" moment when she realized that her work as a literacy and learning instructor in northern Indigenous communities was perpetuating colonial processes similar to those experienced by her own people in Zimbabwe:

> I realized that it [was]…a hand-out type of development, it's not really from within… So I stopped doing those things…It stopped being about Zimbabwean experiences and Canadian experiences. It's the same…When I try to reconcile everything, I look at the experiences in my country where I know how to grow [and cultivate] crops… because we learned those things in geography [and at home]…Everybody knows those things, everybody farms. And so, I know how to take care of the land, but it's very political too because now we are forced to plant all of these…genetically modified crops that come in as aid and that don't do well over time. We don't know how to farm those things and then when [we] reject [it, then it creates major international tension]…You're trying to be self-determining and our politicians who reject this harmful "aid" are framed as monsters depriving their people of food—when you're just trying to feed yourself in a sustainable [and traditional] way.

Takwana's comments are reminiscent of Maori scholar Smith's (1999) discussions of internally driven socioeconomic development as a key principle of Kaupapa Maori, an influential Indigenous-centred pedagogical and community development theory. Her insights also allude to one of the primary societal implications of this study that is discussed in the following chapters, the consideration of "wicked problems" that require interdisciplinary and intercultural collaboration.

Notes

1. Kathy and Takwana elected to use their real names, while Sophia selected her own pseudonym.
2. A grassroots Indigenous and allied movement (Kino-nda-niimi Collective, 2014; Lowan-Trudeau, 2013b).

. 7 .

IMPLICATIONS FOR ENVIRONMENTAL EDUCATION IN CANADA AND BEYOND

Throughout both my doctoral and the subsequent pilot study I questioned how approaches such as ecological métissage, Two- and Three-Eyed Seeing, Integrative Science, and other similar concepts might reshape environmental education in Canada and beyond. My initial response is that it has already begun. For example, along with the growth of intercultural environmental education programs such as those described in Chapter Four, a number of conferences and journal issues (e.g., Korteweg & Russell, 2012; Tuck, McKenzie, & McCoy, 2014; Wildcat, McDonald, Irlbacher-Fox, & Coulthard, 2014) over the past few years have focused on the concepts examined in this study. These kinds of developments provide me with great hope for the future.

So, based on this growth and the experiences and perspectives shared by the participants in these two related studies, how might we be re-shaping environmental education in Canada? Also, what effect might our efforts have on Canadian society in general? In the following I address these questions through discussion of key findings from these two studies. Topics discussed include moving from abstract notions of Aboriginal peoples to authentic engagement and partnerships, re-imagining student-teacher relationships, Canadian cultural and ecological identity, and the role of educators in addressing "wicked" contemporary socio-ecological problems (Vink, Dewulf, & Termeer, 2013).

From Abstract Notions to Authentic Engagement

I was very inspired by one trend that emerged in interviews with several of the Euro-Canadian participants in both my doctoral and follow-up pilot studies—a growth from curiosity and abstract notions of Indigenous peoples and cultures at young ages to authentic engagement later in life due to personal motivation and opportunities presented through school, work, and other involvements.

For example, Skywalker shared early memories of time spent at a family cottage:

> I can remember for sure, 12, maybe younger, when we would drive there, I would look out the window and I always saw myself as this Aboriginal person running in the bush beside the car and if there was a river then I was on the river in a voyageur canoe. You know…I don't know if I'd even heard of a voyageur canoe in a book or anything…but I knew I'd been in a voyageur canoe…It was like…this internal knowledge. I knew… that that was part of who I was on some level… and so…as I got older…it was almost, this would sound corny, but it's almost like the past life part of me was still present and hadn't finished something in this life.

As was presented in Chapter Five, through her work and involvement with other organizations later in life, Skywalker received and created several opportunities to engage for extended periods of time with Aboriginal peoples. Through building reciprocal relationships and being open to learning and making mistakes, her understanding of Aboriginal cultures has deepened significantly since childhood, resulting in an authentic, deeply respectful, and embodied understanding today.

Orange Blossom shared a similar perspective. Discussion of her childhood revealed an innate curiosity about Aboriginal peoples and cultures. She reflected:

> I have no idea why, because growing up we never knew any Native people. There just weren't any where I grew up…But for some reason when I was [young], I read every single book on [Aboriginal people] that I could find in any library…I have had Native people tell me when I tell them that, "Well that's because you were a Native in another life." Hahaha!

Similar to Skywalker, Orange Blossom had the opportunity later in life, in the earlier stages of her teaching career, to work in an Aboriginal community. This experience "opened her eyes" to the realities of contemporary Aboriginal issues and cultures and motivated her to continue working in Aboriginal contexts. As she said herself:

> That course really opened my eyes to...environmental problems...on the reserve...That was the first experience I'd had working in a Native community and...I learned a lot about Native ways of thinking...about the environment...It was like living on another planet for me...But...I just felt from day one that they accepted me. There was not a single negative word that anybody uttered even in the whole community...

Reflecting on how these kinds of experiences have influenced her pedagogical practice, Orange Blossom related the creation of a cultural and ecological field school that brought together Aboriginal and non-Aboriginal students in Haida Gwaii where local Elders and Euro-Canadian academics all contributed to a Third Space educational experience. These kinds of experiences not only deepen the awareness of non-Aboriginal students, but also challenge the diasporic experiences of Aboriginal students (Roth, 2008). By facilitating direct interaction and relationship-building between Aboriginal and non-Aboriginal students in a reciprocal intercultural learning environment, they learn from and about each other in very authentic ways. Aboriginal "voice" (Graveline, 1998) is also introduced into mainstream discourse, challenging the preconceptions and assumptions of non-Aboriginal students and encouraging all students to consider the similarities and differences between Aboriginal and Western epistemological and ontological traditions, another important implication of these kinds of initiatives.

Takwana, a participant in the follow-up study, expressed similar sentiments when asked about the role of educators in helping newcomers to authentically engage with Indigenous knowledge in such complex contexts; she described the importance of basic knowledge and community involvement:

> I think, the general problem is the lack of knowledge that people have...people just don't know. So, just taking the time to learn about their students and just to read...one book...so you get at least some idea of where people are coming from, so you're not just going off of popular rhetoric or what the news is saying...Or, if you're invited to go to something, just go...for one day even, because it changes how you view people.

Takwana continued by connecting her thoughts to a broader societal context and the similarities between Indigenous and immigrant experiences of stereotypes and prejudice:

> Even with Idle No More, if you're going just on what the papers are saying, of course you're not going to have the right frame of mind...The same goes for immigrants, if you go by what the Minister of Immigration is saying, that we're taking all the jobs, of course you're going to have a negative perception of immigrants, but just take the time to learn about us, you know. It doesn't take much, really, just take the time to talk to someone and realize that this is a human being. They're not trying to cause any harm. For many immigrants, going back home will never happen. It's not like we're here because we are trying to take over. We're here out of desperation most times. Some people are running away from wars, you know, there are those stories that people will tell you and you'll realize that these are people like you, you could be in that situation one day. And that's the thing with the reserves...when you go over there, you're treated like a person, you know, they don't reject, and they extend courtesy...

As Naess and Rothenberg (1990) suggest, challenging the reductionist assumptions of Western science and other cultural narratives and encouraging students to consider other ways of understanding and living in the world might also help them to deepen their ecological views, promoting a heightened sense of connection to their home regions. As Tong also noted in my doctoral study, critically informed intercultural engagement can also motivate non-Indigenous students to rediscover their own cultural traditions and to consider how they might incorporate those into their personal and professional lives in the future.

Embracing Pedagogical Compexity

One of the primary implications of these two studies is for educators to re-imagine cultural and pedagogical complexity as a possibility and strength, rather than a challenge or deficit. While some science and environmental educators may remain reticent to foster socio-critical and interdisciplinary dialogue (Chambers, 2011; Steele, 2011) or perhaps feel that they do not have the curricular or logistical space to do so, there is increasing curricular and administrative support for such approaches (Elliot, 2011). Indeed, in some Canadian provinces such as Ontario, interdisciplinary high school programs that bring together the arts, humanities, and sciences have flourished (Breunig, Murtell, Russell, & Howard, 2014; Sharpe & Breunig, 2009).

However, as Elliot notes, educators teaching discrete courses can still do much to foster effective learning through experiential, community-connected, and socio-critical discussions and experiences.

As participants in both studies emphasized, facilitating socio-critical discussions and real-life experiences is a key element for successfully introducing students of all backgrounds to Indigenous knowledge and traditions in Western science settings (Elliot, 2011; Lowan-Trudeau, 2014; Simpson, 2002. As Takwana indicated in the follow-up study, learners new to Canada and their peers will benefit even more when provided with the opportunity to reflect upon and contribute their own culturally based understandings to critical discussions of Western, Indigenous, and other knowledge systems. In this manner, collectively we may well move over time from bioregional bricolage to métissage, from using only one or two eyes, to a dynamic Three- or Multiple-Eyed Seeing model.

A'o/Ako: Re-Imagining Teacher-Student Relationships

Another related meta-dialogue that caught my attention was the participants' descriptions of their pedagogical approaches and philosophies. Participants in both studies emphasized the power of teaching philosophies and practices that include concepts such as humility, building reciprocal relationships, having a sense of humour, and being open to learning *from* and *with* your students. I wouldn't venture so far as to say that these characteristics are *unique* to intercultural environmental educators, but I would suggest that these kinds of beliefs and practices are important *foundations* of effective intercultural environmental education.

Tong, a Euro-Canadian, eloquently described the transformational power of being "comfortable with the uncomfortable" in intercultural situations; she related the challenge and importance of being brave in the face of intercultural conflicts. She described facilitating heated and emotional discussions that touched on challenging topics, but eventually deepened relationships between instructors and students as well as amongst the students themselves.

When asked to share some of her epiphanic teaching moments, Skywalker also expressed this concept. At one point she commented on her own teaching philosophy:

> Another journey I'm on personally is that…You know there's this paradigm where you know everything [as a teacher. However] in this last year I feel like that's been part of my opportunity and challenge is to shut up more and, you know, let the kids… teach me something whether I know it or not…

Skywalker also described several instances where having a sense of humour about herself has allowed her to overcome cultural gaffes and deepen relationships with Aboriginal people. One story that she shared involved a discussion with an Elder about the proper attire for a Sweatlodge. Another insight that she related involved accepting corrective advice from an Aboriginal student on the protocol for conducting a smudging ceremony,[1] an excellent illustration of being humble and open to learning from your students.

A foundational concept in many Indigenous cultures that was recognized by many participants is the importance of reciprocal relationship-building. Orange Blossom expressed this concept eloquently, along with demonstrating her sense of humour, when she described the "sex lives of seashore animals" walking workshops that she would lead in the remote coastal community where she conducted her doctoral research, giving back to the community that was providing her with knowledge of their own.

In the follow-up pilot study, Sophia expressed a similar perspective to Orange Blossom regarding how she might use such experiences to further this dialogue in her work with other researchers and students:

> As I'm building these relationships and building this history of working in the Arctic…I want to use that [to help] other students have those experiences, to foster those relationships. Because…I [was just] thrown [into it], but it's so valuable to [also] have those fostered and…guided experiences…Otherwise…you can't learn as much and you can make a lot more mistakes…

N. Wilson (2008), a Hawaiian scholar and educator, describes this two-way exchange of teaching and learning as "A'o/Ako". I have shared this concept with my own teacher education students by including Wilson's article in course reading lists and have been amazed by their enthusiastic responses; students consistently express their admiration for it in class as well as in response papers and other assignments. Many comment that they have not been exposed to similar approaches in other classes.

In a similar spirit, Arthur expanded the discussion of teacher-learner relationships to the greater-than-human world in my doctoral study when he described his playful, yet earnest, approach to challenging peoples' understanding of themselves through interactions with his team of sled dogs when he stated:

> I…often ask people…to tell me about themselves…and [then] I give them some dogs…Some of the dogs will be much like themselves, but…I will also give them dogs who are…not like themselves, [so] there'll be some dogs that…you're going to have to adjust yourself to in order to make that relationship stronger. And so…I'll often ask them, "What was it like being able to reconcile those differences?" And… I try to engage people in some social learning [focusing on] how they relate to…other people, other beings, other animals, and to the environment around them…They're not expecting to learn those things about themselves sometimes.

Educators who view themselves and their students through a lens of A'o/Ako (N. Wilson, 2008) with humility and humor demonstrate genuine respect for other people, creatures, and cultures. My hope is that, as intercultural environmental education continues to grow and further infiltrates the fields of environmental education and education in general, an increasing number of educators will also embrace the spirit of A'o/Ako.

Canadian Ecological Identity: A Métis Nation?

Another topic of discussion that elicited a host of intriguing insights in both studies was the relationship between Canadian cultural and ecological identity. Saul's (2008) concept of Canada being a "métis" nation that has forgotten and/or ignored the contributions of Aboriginal peoples was also raised as a point of discussion in several interviews. I believe that these concepts are strongly related to intercultural environmental education in Canada, so I explore them in the following as they relate to the growth of our field.

When asked if they agreed with Saul (2008) that Canada is a "métis" nation, the participants provided a variety of responses; some expressed strong agreement while others had sharp criticisms and constructive critique of his perspective. For example, Skywalker expressed strong support for Saul's (2008) concept when she stated:

> I think it's brilliant actually…I mean, ultimately I think it's kind of our way through… Because…it creates that opportunity for us to see that we are all the same in terms of our needs, wants, human nature, makeup, and that we all have capabilities and failings… We need to get beyond our ethnic differences…

Thor was also enthusiastic:

> I think that in many ways we could consider Canada a métis nation. That Canada emerged and developed with the myriad interactions and influences from the Inuit,

> Métis and First Nations peoples is cause enough for me to consider further the concept of a métis nation.

Arthur also added some interesting observations:

> I can't categorize Canada at all culturally, as a whole...It is so diverse...But...what resonates [for me] is...this idea of...movement and sharing of ideas. If you take the word "métis" [to] mean "cross-cultural"...not necessarily Métis as [in] Louis Riel...I think that Canada is very much a métis nation...

However other participants expressed some reticence, such as Liatrus who was hesitant to embrace the concept of Canada being a "métis" nation due to concerns about the "melting pot" effect that results in the loss of unique cultures, but she also expressed some support for Saul's (2008) concept because it encourages people to think critically about Canadian culture and history:

> I can understand it, I can appreciate it, but I think it's maybe...just sort of feeding into the idea of th[e] melting pot...But to say that we're...a métis nation, I think that each of us has our own sort of uniqueness in that, so I think...it's maybe a little bit over simplified. But the concept in and of itself I think is great, because it gets people to actually think, you know? And really think about the history of Canada and all of the interesting dynamics that make this country what it is...

Kate also shared the following comments on how Saul's (2008) concept reminds her of the importance of recognizing the historical and contemporary relationships between Aboriginal and non-Aboriginal peoples in Canada. She also questioned how these concepts can be introduced into pedagogical practice:

> It does speak to me and...I think it's about honouring and recognizing...contributions to our contemporary society and to our history that have been...overlooked and diminished and downplayed...I'm really interested in how we then take his ideas and work with that amongst student groups who haven't even thought of the concept of identity or concept of culture [before]...People tend to put themselves in this unapologetic stance of ignorance where they pretend like they've had nothing to do with Aboriginal peoples or communities, so they don't think they have to engage... But that fails to recognize that...all people in Canada are living in relation to...Aboriginal peoples...Saul is pointing out that...there is this relationship...and we just haven't paid attention to it and here's...the evidence...for why we should.

Cedar Basket, a Sto:lo educator, expressed enthusiasm for Saul's perspective. She also emphasized the historical timeline of the rise of métissage during the fur trade, the decline of respect for Aboriginal peoples in the late nineteenth and early twentieth centuries with increased European settlement, and the contemporary promise of renewed transcultural collaboration:

> I think he's right on…It [is] an invitation to…pause for a moment [and] reflect on…when this Métis idea came alive in our country and…how it grew, and how it got nurtured and how it got damaged sometimes and how it's still very present today…I think…that is really worth exploring…in many different disciplines… Absolutely…We've got to get all Canadian citizens in [the same] canoe…going in the right direction!

Cedar Basket and Kate outlined the historical aspects of Aboriginal and non-Aboriginal relationships in Canada; like Saul (2008), they remind us that, in the early days of exploration and settlement, Europeans relied heavily upon Aboriginal peoples as guides, trading partners, and often as family members. This is not to dismiss the often patronizing or condescending attitudes that accompanied the colonial mentality, but to acknowledge that there were many instances of intercultural cooperation and métissage as I explored in Chapter Three. As Saul (2008) notes, these relationships changed drastically with the influx of greater numbers of European settlers, especially women, resulting in increased discrimination towards and ignorance of Aboriginal peoples. Attitudes of prejudice and cultural superiority continued well into the twentieth century, and, unfortunately, into the present day. However, I am inspired by the growth of areas such as intercultural environmental education as signs that these relationships are changing once again. As so many of the participants in both studies expressed, it is still possible to forge and renew relationships built on intercultural and, eventually, transcultural cooperation.

Exploring Canadian Ecological Identity: A Varied Terrain

Another topic that produced several interesting conversations was Canadian ecological identity. It quickly became apparent to me that no single Canadian ecological identity can be identified. Participants in both studies expressed a variety of opinions on the origins and expressions of the various ecological identities that they identified within and across Canada. Influences identified ranged from geography and history to socioeconomics and culture. The expression of these identities was also described in different ways—through exploration of political, recreational, occupational, and economic perspectives and practices.

Arthur identified variations in ecological identity arising from different subcultures and regions in Canada:

> I don't think there is one ecological identity…I think that [it] is probably different in different [sub-]cultures…So [in] business culture, our ecological identity is one of natural resources…Internationally as well, our ecological identity is [linked to the] business world [and resource extraction] I think that probably shapes our… ecological identity a little bit more than anything else…But I think that culturally…the importance of the symbol of the beaver to Canada…and other things…like maple syrup [and]…the maple leaf [is] funny because…that's not my experience of Canada… There [is so much regional and] ecological diversity…Canada's not all maple trees…

Thor also commented on individual and regional variations of ecological identity in Canada:

> Within…any family, community, or province, I don't think there's a common ecological identity…However, in the community where I was living…in Nunavut last year… there amongst the locals, I think more so than any one place I've been, there was more of a common ecological identity…which is very different from the ecological identity of…the general mass of people in southern Ontario where I grew up…

As Arthur and Thor discuss above, geography has been a very powerful influence in the past and present on Canadians' ecological identities. This concept lends support to the previously discussed concept of bioregionalism and deep ecology's emphasis on locally developed values and practices. Similar to Aboriginal scholars such as Cajete (1994), Thor also commented on the historical connection between both Aboriginal and more recent settler cultures and the geography of North America:

> The rivers and the waterways being traditional and modern travel routes…have influenced the location of most if not all the major cities in Canada…Further, the waterways, the forests, the mountains and everything that comprises the landscape is inherently connected to the Aboriginal cultures that have lived in these parts for… some say time immemorial…thousands of years at least…

Arthur also discussed the (often forgotten) legacy of the fur trade and the influence of Aboriginal peoples on the building of Canada as we know it today. Similar to Dean (2006) and Newbery (2012), he also commented on the ignorance of many Canadians regarding the origins of the canoe:

> The fur trade is when ideas moved and when people really started interacting with…different cultures…and sharing things…When you think about Canada's…

> geography...the shape of the land...[and] the Great Lakes...right in the middle of the continent...it makes so much sense, it's not random, we are shaped by our environment...So many Canadians...celebrate the canoe as [a] wonder of Canada...But it's interesting because the ones who are doing the canoeing [often don't have] any clue [why the canoe is important] or why...the centres of business [are] where they are...And so...that started out of necessity...and became a recreational activity... [However] I think that our identity ecologically...has...been disconnected [from that history].

As Thor and Arthur emphasized, Canada's geography has drastically influenced the formation, movement, and settlement of people since time immemorial. Like Saul (2008), they both also highlighted that the foundational role of Aboriginal peoples in the formation of our contemporary nation has been largely forgotten and/or ignored. This is another area where intercultural environmental educators can raise awareness, providing their students with a deeper and more authentic understanding of the past to inspire the future.

Similar to Arthur and Thor, Liatrus emphasized the diversity of ecological identities in Canada. She also described what she perceived as a gradual shift towards greater ecological empathy due to factors such as the growth of environmental education and increased understanding of the relationship between our cultural and ecological identities:

> I would say that it would be so diverse [that] to try to describe it as a universal...is a mistake...It's also evolving... I mean we think about environmental education and... changing dynamics and all the people like you and I...So...it is so mixed that...even that notion of...a mixed cultural [and] ecological identity may be part of the definition of what [it] is...

Kate also described the various subcultures that she believes have a strong influence on individuals' ecological identities:

> It might be a stretch to say that we have one national Canadian ecological identity...I think there are...those people that envision their...ecological connections or contributions to be quite urban and...sometimes also intertwined with...social issues...I think there are other groups...who don't necessarily pay attention to social issues and who are more..."Back to Nature"...I think there's people whose ecological identity...is really fueled by a physicality—they're seeking adventure and sport... And I think there are, you know, people who are quite spiritually connected to Nature and...people who are ancestrally connected to particular localized places...Then I also think there's people who don't think about it all...who are so far removed that it makes them nervous to see a spider or...[who] spend their whole life living in their air-conditioned house...car [and] office.

Natasha's comments were reminiscent of Francis (2005) who discusses the cultural importance of symbols such as the canoe and the North to Canadian identity. She also anticipated the previously presented perspectives of participants such as Takwana and Kathy in the follow-up pilot study when she noted Canada's increasing cultural diversity and questioned if the experiences of more recent immigrants would be different than her own:

> Canada as a whole has an ecological identity just in terms of…the way that people identify…with…[the] outdoors…[For example] there's a canoeing culture in Canada. Even if people don't canoe, they still recognize the canoe as a Canadian symbol… The North, you know, thinking of Inukshuks and…the northern landscape is very much a part of a cultural landscape in Canada…Or open prairie landscapes, you know, this is part of my [own] identity as a Canadian and I've never been to the prairies. So…I think that there's that kind of…landscape within the identity…But, at the same time, there are so many cultures within [Canada]. I mean…it's very much a blend of cultures…and…I don't know how others would feel…you know…first generation descendants of immigrants to Canada…I mean, having been raised in Canada and my parents and grandparents and great grandparents and great-great grandparents…being born in Canada, I think probably gives me a different perspective and kind of a different claim over "Canada"…and also being raised in a more rural area…I definitely think that there's a Canadian claim to Nature…which would feed into our collective ecological identity.

Cedar Basket optimistically described a shift towards greater ecological awareness that she has witnessed in family members who work in forestry:

> [Based on] thirty years of conversations with loggers in my family…what I am hearing is that there is concern about how the practices of the past cannot continue into the future. And so, I sense a shift, how[ever] subtle…just within my family…sphere… There is concern that things are connected. [If] you clear cut a whole range of forest…the soil tumbles down into the salmon beds and they can't spawn anymore… It's like a domino [effect and] people are seeing this. The things that they used to see…in childhood aren't there anymore or they hardly exist…Things have to be done differently. And so…I want to believe that Canada is moving…more…into a concern for…environment…Everything comes around in cycles, [so] I may be just old enough to hopefully see the next big positive push for a nation-wide change in our environment policy to the better. Where we're looking at growing…instead of monopolizing…That's just one River Woman's hope…

Cedar Basket's optimism is inspiring. Like her, I hope that we are moving towards a deeper collective ecological awareness that transcends the political, cultural, and socioeconomic divisions described by the other participants above. Cedar Basket's comments also allude to the interdisciplinary nature of "wicked problems" and the transdisciplinary, transcultural strategies that they require.

Note

1. A smudge is a purification ceremony that involves the burning of herbs, also called sacred medicines, such as sweetgrass, sage, and cedar to produce a cleansing smoke (Portman & Garrett, 2006). The smoke from a smudge may be used to spiritually cleanse people, places, and objects.

. 8 .

FINAL THOUGHTS & FUTURE DIRECTIONS

As I came to the end of this research journey, I found myself pondering its significance; how does it relate to and/or depart from the literature, concepts, and participants perspectives presented in previous chapters? How is it unique? And what have I found or experienced that contributes fresh perspectives, understanding, and new directions to the field of intercultural environmental education?

Engaging Wicked Problems through Education

Vink, Dewulf and Termeer (2013) define "wicked problems" as those "which cannot be precisely formulated or solved, because of widely diverging problem formulations and vested interests" (p. 45). They note that considering and addressing wicked problems such as climate change adaption is one of the most pressing and confounding challenges of our time.

Pedagogical models that promote the participation of multiple stakeholders to consider and address wicked environmental problems, such as Two- or Three-Eyed Seeing and ecological métissage, hold great promise as they allow for the contribution of Western, Indigenous, and other culturally rooted understandings from around the globe. As science and environmental educators, we can support this shift away from decision-making based exclusively on Western science by fostering critical and collaborative pedagogical settings that honour and welcome the culturally rooted knowledge and perspectives of all of our students (Carter, 2011; Ferkany & Whyte, 2012).

Ensuring exposure, links to, and critical consideration of contemporary wicked problems such as climate change and current disputes over Indigenous land rights and resource development across Canada and, indeed, around the world, is also crucially important. We must assist our students to develop the skills required to critically draw from the best of all cultures and knowledge systems as appropriate while avoiding the homogenizing pitfalls of a transcultural globalized melting pot scenario (Carter, 2011; Welsch, 1999).

My optimistic hope is that such a shift in not only science and environmental education, but society in general, will nurture a future generation of leaders who truly value ecological knowledge and values other than those exclusively based on Western science. How much further ahead would we be in considering and addressing wicked problems such as climate change, widespread fisheries and wildlife depletion, seasonal flooding, and other highly complex socio-ecological problems if the wisdom and knowledge of Indigenous and other non-Western peoples had been equitably included from the start? As Sophia noted during our conversation for the pilot study:

> I think that as much as academically we talk more about local and Indigenous knowledge, I think that in broader society...scientists [are still] put...on a pedestal...[We need to] cultivate more respect for that local knowledge in ecosystem management [and] industrial development projects.

I believe that we are slowly seeing increasing inclusion of Indigenous perspectives in tackling socio-ecological issues in Canada and other nations. Increasing co-management of ecosystems (Menzies, 2006) and promising examples such as the recent Cohen Commission that was mentioned by Cedar Basket during our interview, an inquiry into the shocking depletion of sockeye salmon stocks in British Columbia's Fraser River that drew on a broad range of stakeholders including representatives from government, industry, Indigenous and non-Indigenous communities, and Western science knowledge holders (Cohen, 2012) are inspiring to be sure. As Kassam (2014) has also noted, dialogue and sharing between Indigenous peoples from similar geographical and ecological areas is also providing further insights to address wicked problems such as climate change in the Arctic.

Recent court decisions also provide me with great hope. For example, shortly after the federal government's controversial approval of the Northern Gateway pipeline (Natural Resources Canada, 2014) that would carry bitumen-laden tar sands oil from Alberta over 1000 kilometres across several mountain ranges and 1000 salmon-bearing waterways to a treacherous stretch of the Pacific Coast (Boulton, 2013; Gunton & Broadbent, 2013), Justice McClachlin's historic judgement (*Tsilhqot'in Nation* v. *British Columbia*, 2014) in favour of the Tsilqhot'in in central British Columbia, affirming their right to much stronger control over resource development and ecosystem management in their traditional territory is extremely uplifting. As environmental educators, we must be aware and facilitate discussion of the impact of such societal dynamics with our students. Exploring the inherent tension within such pedagogical endeavours is another line of inquiry that I am currently pursuing.

Intercultural Alliances

Another common topic of discussion in both studies was the importance of building strong intercultural alliances that acknowledge and incorporate multiple cultural perspectives in authentic ways. In different ways, all of the participants supported such an approach in order to honour the individual and contextual perspectives of both Indigenous peoples and newcomers to Canada in the spirit of living well together on this land (Haluza-Delay, DeMoor & Peet, 2013; Root & Dannenman, 2009; Scully, 2012). However, the role of non-Indigenous allies in Indigenous environmental education and activism is an area that remains to be adequately explored in the literature. While a growing number of authors have recently made valuable contributions and raised important questions and concerns regarding the role of non-Indigenous allies in Indigenous education and activism in Canada and elsewhere (e.g. Davis, 2010; Korteweg & Oakley, 2014; Root, 2010; Scully, 2012), there is still space to expand such considerations and their application in practice.

For example, as Sophia emphasized in the pilot study, allied activism against the proposed Northern Gateway pipeline in Western Canada is a case in point exemplifying intercultural cooperation and respect for Indigenous knowledge:

> People [are] standing [together] and saying, "No…[We] can be White or Indigenous or [Métis] and we can stand together…We can support each other…And I think that strength is building and that's the environmental ethic we need.

Inspired by such statements and based on my own experiences as an activist, I am currently exploring the pedagogical potential of ecological activism in Indigenous contexts in another study. That study also examines the experiences of Indigenous and non-Indigenous educators regarding the introduction of such discussions in educational settings (Lowan-Trudeau, 2015) and seeks to make more explicit connections with and gain insights from related fields such as environmental justice (Agyeman, Cole, Haluza-DeLay & O'Riley, 2009; Russell & Fawcett, 2013) and ecofeminism (Martusewicz, 2013).

Future Theoretical and Educational Program Development

I am also looking forward to applying the findings from these two studies study to guide the development of an action research partnership with a community-based science and environmental education program interested in emphasizing and integrating Indigenous ecological knowledge and philosophy in science and environmental education. Further inquiry into the experiences of youth and adult learners, and educators engaging with these complex situations will most certainly prove insightful and further the consideration, conceptual development, and application of conceptual models such as ecological métissage and Three-Eyed Seeing.

Further Research Possibilities

These studies were limited to the Canadian context. However, I did draw on international literature to expand and deepen my understanding of concepts such as métissage, and Indigenous and Métis experiences and identities around the world. These explorations led me to understand that there are certain commonalities of experience and perspectives emerging from those nations influenced by the ongoing legacy of colonialism. Thus, I encourage those who read this from beyond the borders of my own country to interpret the themes and concepts presented through their own sociocultural and ecological lenses.

Another limitation of these two studies from a Canadian perspective is that, while some supportive French literature was accessed, all of the interviews were conducted in English with primarily Anglophone participants. As such, I agree with Russell and Fawcett (2013) who suggest that the monolingual proclivity of our field is a limitation that deserves greater attention. For example, during recent participation in several national and international environmental education conferences I observed that, while excellent research and pedagogy is happening in many areas of the world, cultural and linguistic barriers sometimes dissuade us from sharing and/or collaborating. This limitation points to another area of possible future research—exploring and strengthening collaboration efforts between the French, English, Indigenous, and other languages of environmental education communities in Canada and internationally.

Significance

Coming to the end of this research journey prompted me to consider its significance to the field of environmental education. One significant contribution is that it adds greater cultural diversity to the body of significant life experience research. As delineated by Dillon et al. (1999), previous significant life experience research has been primarily conducted from a Eurocentric perspective examining Eurocentric approaches to environmental education. I believe that these studies have added other perspectives to this dialogue through the voices of participants who are of Sto:lo, Métis, Pakistani, Shona, and Japanese ancestry. A diversity of perspectives were also presented by the Euro-Canadian participants who all expressed a deep understanding of and respect for Indigenous cultures and articulated their own unique identities, philosophies, and practices.

I believe that the findings from these studies also challenge the persistently perceived dichotomy between Western and Indigenous knowledge and philosophies of Nature. Similar to Tong and Skywalker, a limited few intercultural environmental education scholars such as Kawagley and Barnhardt (1999) have acknowledged the existence of Western traditions other than science, although none have expanded upon their relationship (potential or embodied) to Indigenous traditions in detail. By explicitly and implicitly acknowledging and exploring the development, influence, and potential of Western-derived perspectives such as deep ecology, bioregionalism, and place-based education, this endeavour has contributed to a softening of the borders between Western and Indigenous approaches to environmental education; many of the educators profiled in this study not only draw on Western science and Indigenous traditions, but are also informed, for example, by deep ecological perspectives.

The development, articulation, and critical consideration of my own ecosophy, ecological métissage, also contributed to this discussion.

Many Western environmental philosophers over the past two centuries have followed "archetypes" (Thomashow, 1996) such as Thoreau, Muir, and Carson in advocating for the consideration of Indigenous perspectives. However, few have demonstrated an authentic and engaged understanding of what this actually entails. I believe that this work has contributed to the field by sharing the stories of environmental educators and leaders, Indigenous and non-Indigenous alike, who are engaged in authentic, locally grounded, and engaged relationships and endeavors.

My research also critically examined and expanded upon Saul's (2008) concept that Canada is a "métis nation". One gap in Saul's discussion was in the area of environmental perspectives; while he did present a short section that advocated for the potential contribution of Indigenous knowledge to contemporary environmental discourse in Canada, he did not expand upon it in any great detail, nor did he examine it through his lens of Canada as a "métis nation". I believe that my work has provided further exploration of this area through a lens of métissage. As described previously, many of the participants agreed with Saul as to the métis foundations of early Canada, but many also challenged his notion that it remains so, emphasizing the progressively oppressive nature of European colonialism in Canada and the attendant cultural and ecological consequences.

Finally, and perhaps, most significantly, I believe that these two studies have helped to further explore and articulate the relationship between Western, Indigenous, and other culturally rooted epistemologies and the use of concepts such as bricolage, métissage, and Two- and Three-Eyed Seeing as metaphoric models for research and pedagogical praxis.

In the second chapter I described my experiences as a methodological métisseur seeking to articulate and enact a métissage of Indigenous and interpretive research methodologies. Realizing that this was possible and that I had achieved it, albeit largely intuitively in the beginning, was one of the most exciting moments of this research journey for me.

Final Thoughts

Through these studies and other work over the past several years, I have had the immense pleasure of meeting and working with intercultural environmental educators from across Canada and around the world, physical and epistemological border crossers working towards increased intercultural and ecological consciousness. Roth (2008) states that the world needs more people such as this who are able to "live and create diasporic identities" (p. 915).

As embodied by the programs, educators, students, and community leaders profiled in this book, the Third Space is once again taking hold across Canada and around the world. Educators committed to such an approach are developing and facilitating programs that foster locally grounded authentic engagement and collaboration between Indigenous and non-Indigenous peoples and epistemologies, conscious of their simultaneous role as teachers and learners, with a sense of humor and humility. I am inspired by the growth of our field; my hope is that, in the decades to come we will experience a continued increase in critically informed intercultural and inter-Indigenous sharing and collaboration in a spirit of mutual respect, reconciliation, and a common love for the Land.

REFERENCES

Aberley, D. (1999). Interpreting bioregionalism: A story from many voices. In M.V. McGinnis (Ed.) *Bioregionalism* (pp. 13–42). New York: Routledge.

Absolon, D., & Willett, C. (2005). Putting ourselves forward: Location in Aboriginal research. In L. Brown, & S. Strega, (Eds.), *Research as resistance: Critical, Indigenous, and anti-oppressive approaches* (pp. 97–126). Toronto: Canadian Scholars' Press.

Adams, H. (1999). *Tortured people: The politics of colonization, the revised edition*. Penticton, B.C.: Theytus Books Ltd.

Adams, J., Luitel, B.C., Alfonso, E., & Taylor, P.C. (2008). A cogenerative inquiry into postcolonial theory to envisage culturally inclusive science education. *Cultural Studies of Science Education*, 3, 999–1019.

Agyeman, J., Cole, P., Haluza-DeLay, R., & O'Riley, P. (Eds.) (2009). *Speaking for ourselves: Environmental justice in Canada*. Vancouver, BC: University of British Columbia Press.

Agyeman, J. (2003). "Under participation" and ethnocentrism in environmental education research: Developing "culturally sensitive research approaches". *Canadian Journal of Environmental Education*, 8(1), 80–94.

Aikenhead, G., & Michell, H. (2011). *Bridging cultures: Indigenous and scientific ways of knowing nature*. Toronto, ON: Pearson.

Angleviel, F. (2008). Trois millénaires de migrations et de métissages en Nouvelle-Calédonie: Réalité biologique et deficit culturel. *International Journal of Francophone Studies, 11*(4), 523–537.

Anishnawbe Health Toronto (2008). *Sweatlodge*. Retrieved March 15th, 2008 from http://www.aht.ca/resources/traditional_teachings/sweat_lodge.

Backhouse, F. (2008, Summer). A Skookum language. *British Columbia Magazine*. Retrieved November 28th, 2008 from http://www.backhouse.ca/a_skookum_language.php.

Bakker, P., & Papen, R.A. (1997). Michif: A mixed language based on Cree and French. In S. Thomason (Ed.) *Contact languages: A wider perspective* (pp. 295–363). Philadelphia: John Benjamins.

Barnhardt, R., & Kawagley, A.O. (2005). Indigenous knowledge systems and Alaska Native ways of knowing. *Anthropology and Education Quarterly*, *36*(1), 8–23.

Bartels, D., & Bartels, A. (2005). Mi'gmaq lives: Aboriginal identity in Newfoundland. In U. Lischke & D.T. McNabb (Eds.), *Walking a tightrope, Aboriginal people and their representations* (pp. 249–280). Waterloo, ON: Wilfrid Laurier Press.

Bartlett, C. (2005). Knowledge inclusivity: "Two-Eyed Seeing" for science for the 21st Century. In M. Wiber & J. Kierney (Eds.) *Learning communities as a tool in natural resource management: Proceedings from a workshop held in Halifax, Nova Scotia*. Retrieved September 18th, 2014 from http://www.integrativescience.ca/uploads/articles/2005November-Bartlett-text-Integrative-Science-Two-Eyed-Seeing-Aboriginal-learning-communities.pdf.

Bastien, B. (2003). The cultural practice of participatory transpersonal visions: An Indigenous perspective. *ReVision*, *26*(2), 41–48.

Battiste, M. (1998). Enabling the Autumn seed: Toward a decolonized approach to Aboriginal knowledge, language, and education. *Canadian Journal of Native Education*, *22*(1), 16–27.

Battiste, M. (2005). You can't be the global doctor if you're the colonial disease. In L. Muzzin, P. Tripp (Eds.), *Teaching as activism, equity meets environmentalism* (pp.121–133). Montreal and Kingston: McGill-Queen's University Press.

Berry, K. (2006). Research as bricolage: Embracing relationality, multiplicity, and complexity. In K. Tobin & J. Kincheloe (Eds.), *Doing educational research* (pp. 87–115). Rotterdam, The Netherlands: Sense Publishers.

Berry, W. (2009). *Bringing it to the table: On farming and food*. Berkeley, CA: Counterpoint Press.

Bhabha, H. (1998). Cultures in between. In D. Bennett (Ed.), *Multicultural states: Rethinking difference and identity* (pp. 29–38). London: Routledge.

Blanchet-Cohen, N., & Reilly, R. (2013). Teachers' perspectives on environmental education in multicultural contexts: Towards culturally-responsive environmental education. *Teaching and Teacher Education*, 36, 12–22.

Bolak, H.C. (1997). Studying one's own in the Middle East: Negotiating gender and self-other dynamics in the field. In Hertz, R. (Ed.), *Reflexivity & voice* (pp. 95–118). Thousand Oaks, CA: Sage Publications.

Bonniol, J.L., & Benoist, J. (1994). Hérédités plurielles: Représentations populaires et conceptions savantes du métissage. Laboratoire d'Écologie Humaine, Université d'Aix Marseille III, France, 1–37.

Boulton, M. (2013). *Financial vulnerability assessment: Who would pay for oil tankers spills associated with the Northern Gateway pipeline?* Victoria, BC: The Environmental Law Centre, University of Victoria.

Brandt, C.B. (2008). Discursive geographies in science: Space, identity, and scientific discourse among Indigenous women in higher education. *Cultural Studies in Science Education, 3*, 703–730.

Breunig, M., Murtell, J., Russell, C., & Howard, R. (2014). The impact of integrated environmental studies programs: Are students motivated to act pro-environmentally? *Environmental Education Research, 20*(3), 372–386.

Brown, J.S. (1983). Women as centre and symbol of the emergence of Metis communities. *Canadian Journal of Native Studies*, *3*(1), 39–46.

Burley, D.V., & Horsfall, G.A. (1989). Vernacular houses and farmsteads of the Canadian Metis. *Journal of Cultural Geography, 10*(1), 19–33.

Cajete, G. (2001). Indigenous education and ecology: Perspectives of an American Indian educator. In J.A. Grim (Ed.) *Indigenous traditions and ecology: The interbeing of cosmology and community*. Cambridge, MA: Harvard University Press.

Cajete, G. (2000). *Native science: Natural laws of interdependence*. Santa Fe, NM: Clear Light Publishers.

Cajete, G. (1999). "Look to the mountain": Reflections on Indigenous ecology. In G. Cajete (Ed.) *A people's ecology: Explorations in sustainable living* (pp. 2–20). Santa Fe, NM: Clearlight Publishers.

Cajete, G. (1994). *Look to the mountain: An ecology of Indigenous education*. Skyland, NC: Kivaki Press.

Campbell, M. (1983). *Half-breed*. Halifax, NS: Goodread Biographies.

Carter, L. (2011). Gathering in threads in the insensible global world: The wicked problem of globalisation and science education. *Cultural Studies of Science Education*, *6*(1), 1–12.

CBC First Nation Unveils Solar Power Project. (2009, July 17). *cbc.ca*. Retrieved July 1st, 2011 from http://www.cbc.ca/news/canada/british-columbia/story/2009/07/17/bc-tsouke-solar-power.html.

Chambers, C. (2002). I was born into a mixed clan. *Educational Insights*, *7*(2). Retrieved April 10th, 2010 from http://www.ccfi.educ.ubc.ca/publication/insights/v07n02/metissage/a_mixed.html.

Chambers, C., Donald, D., & Hasebe-Ludt, E. (2002). Creating a curriculum of métissage. *Insights*, *7*(2). Retrieved February 2nd, 2010 from http://ccfi.educ.ubc.ca/publication/insights/v07n02/metissage/metiscript.html.

Chambers, J. (2011). Right time, wrong place? Teaching about climate change in Alberta schools. *Alberta Science Education Journal*, *42*(1), 4–12.

Choo, C. (2007). Eurasians: Celebrating survival. *Journal of Intercultural Studies*, *28*(1), 129–141.

Clandinin, D.J., & Connelly, F.M. (2000). *Narrative inquiry: Experience and story in qualitative research*. San Francisco: Jossey-Bass Publishers.

Climate Action Network. (2013). *Canada wins "Lifetime Unachievement" fossil award at Warsaw climate talks*. Retrieved September 18th, 2014 from http://climateactionnetwork.ca/2013/11/22/canada-wins-lifetime-unachievement-fossil-award-at-warsaw-climate-talks/#sthash.3sZlcshi.dpuf.

Coates, J., Gray, M., & Hetherington, T. (2006). An "ecospiritual" perspective: Finally, a place for Indigenous approaches. *British Journal of Social Work, 36*, 381–399.

Cohen, B.I. (2012). The uncertain future of Fraser River sockeye: Volume 1: The sockeye fishery. Ottawa, ON: Minister of Public Work and Government Services Canada.

Cole, P. (2002). Aboriginalizing methodology: Considering the canoe. *International Journal of Qualitative Studies in Education, 15*(4), 447–459.

Crawford, J.C. (1985). Speaking Michif in four Métis communities. *The Canadian Journal of Native Studies, 3*(1), 47–55.

Creswell, J.W., & Miller, D.L. (2000). Determining validity in qualitative inquiry. *Theory Into Practice, 39*(3), 124–130.

Creswell, J.W. (2002). *Educational research: Planning conducting and evaluating quantitative and qualitative Research.* Upper Saddle River, NJ: Pearson Education Inc.

Curthoys, L.P. (2007). Finding a place of one's own. *Canadian Journal of Environmental Education, 12*(1), 68–79.

Cuthbertson, B., Heine, M., & Whitson, D. (1997). Producing meaning through movement: An alternative view of sense of place. *The Trumpeter: Journal of Ecosophy, 14*(2), 72–75.

Csikszentmihalyi, M. (1990). *Flow: The psychology of optimal experience.* New York: Harper and Row.

Davis, L. (2010). Introduction. In L. Davis (Ed.), *Alliances: Re/Envisioning Indigenous-non-Indigenous relationships* (pp. 1–12). Toronto: University of Toronto Press.

Dean, M. (2006). The centennial voyageur canoe pageant as historical re-enactment. *Journal of Canadian Studies, 40*(3), 43–67.

Dean, M. (2013). *Inheriting a canoe paddle.* Toronto: University of Toronto Press.

Denzin, N. (1989). *Interpretive biography.* Thousand Oaks, CA: Sage Publications.

Détienne, M., & Vernant, J.P. (1991). *Cunning Intelligence in Greek Culture and Society.* Trans. Janet Lloyd. Chicago: U of Chicago Press.

Détienne, M., & Vernant, J.P. (1974). *Les ruses de l'intelligence: La mêtis des Grecs.* Paris: Flammarion.

Devall, B. (1988). *Simple in means, rich in ends: Practicing deep ecology.* Salt Lake City, UT: Gibbs Smith.

Devine, H. (2010). Being and becoming Métis: A personal reflection. In C. Podruchny & L. Peers (Eds.), *Gathering places: Aboriginal and fur trade histories* (pp. 181–210). Vancouver, BC: UBC Press.

Dillon, J., Kelsey, E., & Duque-Aristizábal, A.M. (1999). Identity and culture: Theorising emergent environmentalism. *Environmental Education Research, 5*(4), 394–405.

Dodge, J. (1981). Living by life. *CoEvolution Quarterly, 32*, 6–12.

Dolmage, J. (2009). Metis, mêtis, mestiza, Medusa: Rhetorical bodies across rhetorical traditions. *Rhetoric Review, 28*(1), 1–28.

Donald, D.T. (2010). Forts, curriculum, and Indigenous metissage: Imagining decolonization of Aboriginal-Canadian relations in educational contexts. *First Nations Perspectives: The Journal of the Manitoba First Nations Education Resource Centre, 2*(1), 1–24.

Donald, D.T. (2009). *The Pedagogy of the fort: Curriculum, Aboriginal-Canadian relations, and Indigenous metissage.* Unpublished PhD dissertation, University of Alberta, Edmonton, AB.

Dorion, L., & Préfontaine, D.R. (1999). Deconstructing Métis historiography: Giving voice to the Métis people. In L.J. Barkwell, L. Dorion, and D.R. Préfontaine (Eds.), *Resources for Métis researchers* (pp. 3–30). Winnipeg, MB: Louis Riel Institute & Saskatoon, SK: Gabriel Dumont Institute.

Drake, L. (2006, August 28th). Patience runs dry over toxic water. *Edmonton Journal*, A1.

Drengson, A. (2008). The life and work of Arne Naess: An appreciative overview. In A. Drengson and B. Devall (Eds.) *Ecology of wisdom: Writings by Arne Naess* (pp. 3–44). Berkeley, CA: Counterpoint.

Durst, D. (2004). *Partnerships with Aboriginal researchers: Hidden pitfalls and cultural pressures.* Paper presented at the Saskatchewan Institute of Public Policy, November 18th.

Duval, J. (2001–2003). The Catholic Church and the formation of Métis identity. *Past Imperfect*, 9, 65–87.

Edge, L., & McCallum, T. (2006). Métis identity: Sharing traditional knowledge and healing practices at Métis elders' gatherings. *Pimatisiwin*, 4(2), 83–116.

Elections Canada (2011). *2011 general election*. Retrieved September 13th, 2011 from http://enr.elections.ca/National_e.aspx.

Elliot, F. (2011). From Indigenous science examples to Indigenous science perspectives. *Alberta Science Education Journal*, *41*(1), 4–10.

Erasmus, P. (1999). *Buffalo days and nights: As told to Henry Thompson*. Calgary, AB: Fifth House Ltd.

Evering, B., & Longboat, D.R. (2013). An introduction to Indigenous environmental studies: From principles into action. In A. Kulnieks, D.R. Longboat & K. Young (Eds.) *Contemporary Studies in Environmental and Indigenous Pedagogies* (pp. 241–258). Rotterdam: Sense.

Ferguson, W. (2007). *Why I hate Canadians*. Vancouver, BC: Douglas & McIntyre.

Ferkany, M., & Whyte, K. (2013). The importance of participatory virtues in the future of environmental education. *Journal of Agricultural & Environmental Ethics*, *25*(3), 419–434.

Finnegan, R. (1996). A note on oral tradition and historical evidence. In D.K. Dunaway & W.K. Baum (Eds.), *Oral history: An interdisciplinary anthology, 2nd Edition* (pp. 126–134). Lanham, MD: Altamira Press. (Original work published 1970).

Foster, J.E. (2007a). The Métis: The people and the term. In P.C. Douad (Ed.), *The western Métis: Profile of a people* (pp. 21–30). Regina, SK: Canadian Plains Research Centre. (Original work published 1978).

Foster, J.E. (2007b). Wintering, the outsider male and the ethnogenesis of the western plains Métis. In P.C. Douad (Ed.), *The western Métis: Profile of a people* (pp. 91–103). Regina, SK: Canadian Plains Research Centre. (Original work published 1994).

Francis, D. (2005). *National dreams: Myth, memory, and Canadian history*. Vancouver, BC: Arsenal Pulp Press.

Fujiwara, A. (2001–2003). Reconsiderations of frameworks of ethnic history: A comparison of Métis and Ukranian-Canadian historiographies. *Past Imperfect*, 9, 43–63.

Gibbs, E.A. (2000). *The changing face of the Metis nation*. Unpublished master's thesis, University of Lethbridge, Lethbridge, AB.

Ginsburg, F. (1997). The case of mistaken identity: Problems in representing women on the right. In Hertz, R. (Ed.), *Reflexivity and voice* (pp. 283–299). Thousand Oaks, CA: Sage Publications.

Grant, J. (2009). *Come thou tortoise*. Toronto, ON: Knopf Canada.

Graveline, F.J. (1998). *Circle works: Transforming Eurocentric consciousness*. Halifax: Fernwood Press.

Grele, R. (1994). History and the languages of history in the oral history interview: Who answers whose questions and why? In E.M. McMahan & K.L. Rogers (Eds.), *Interactive oral history interviewing* (pp. 1–18). Hillsdale, NJ: Lawrence Earlbaum Associates.

Grimwood, B., Haberer, A., & Legault, M. (2014). Guides to sustainable connections? Exploring human-nature relationships among wilderness travel leaders. *Journal of Adventure Education and Outdoor Learning*. DOI: 10.1080/14729679.2013.867814.

Gruzinski, S. (2004). Occidentalisation, globalisation et métissage dans les Amériques ibériques. *L'Experience Métisse*, 76–83.

Guilloux, D. (2007). *Paddling, portaging and pageantry*. Rocky Mountain House, AB: Doreen Guilloux Publishing.

Gunton, T., & Broadbent, S. (2013). *A spill risk assessment of the Enbridge Northern Gateway project*. Burnaby, BC: School of Resource and EnvironmentalManagement, Simon Fraser University.

Haluza-Delay, R., DeMoor, M.J., & Peet, C. (2013). That we may live well together on this land: Place pluralism and just sustainability in Canadian and environmental studies. *Journal of Canadian Studies/Revue d'études Canadiennes*, *47*(3), 226–256.

Hanrahan, M. (2000). Industrialization and politicization of health in Labrador Métis society. *The Canadian Journal of Native Studies*, *20*(2), 231–250.

Hanson, J., & Kurtz, D.V. (2007). Ethnogenesis, imperial acculturation on the frontiers, and the production of ethnic identity: The Genizaro of New Mexico and the Red River Métis. *Social Evolution and History*, 6(1), 3–37.

Hart, P. (2002). Narrative, knowing, and emerging methodologies in environmental education research: Issues of quality. *Canadian Journal of Environmental Education*, *7*(2), 140–165.

Hasebe-Ludt, E., Chambers, C.M., & Leggo, C. (2009). Life writing and literary métissage as an ethos for our times. New York: Peter Lang.

Hatcher, A., Bartlett, C., Marshall, M., & Marshall, A. (2009). Two-eyed seeing: A cross-cultural science journey. *Green Teacher*, 86, 3–6.

Hatcher, A., & Bartlett, C. (2009a). MSIT: Transdisciplinary, cross-cultural science. *Green Teacher*, 86, 7–10.

Hatcher, A., & Bartlett, C. (2009b). Traditional medicines: How much is enough? *Green Teacher*, 86, 11–13.

Henderson, C. (2013). *Aboriginal power: Clean energy and the future of Canada's First Peoples*. Erin, ON: Rainforest Editions.

Henley, T. (1989). *Rediscovery: Ancient pathways, new directions: Outdoor activities based on native traditions*. Edmonton, AB: Lone Pine Publishing.

Hermes, M. (2000). The scientific method, Nintendo, and eagle feathers: Rethinking the meaning of "culture based" curriculum at an Ojibwe tribal school. *Qualitative Studies in Education, 13*(4), 387–400.

Hogue, M. M. (2012). Interconnecting Western and Aboriginal paradigms in post-secondary science education: An action research approach. *Journal of the Canadian Association for Curriculum Studies, 10*(1), 77–114.

Hones, D.F. (1999). Making peace: A narrative study of a bilingual liason, a school, and a community. *Teachers College Record, 101*(1), 106–134.

Integrative Science Institute (2012). *Bringing together Indigenous ways of knowing and Western scientific knowledge.* Retrieved March 18th, 2015 from http://www.integrativescience.ca

Karahasan, D. (2008). *Métissage in New France: Frenchification, mixed marriages, and Métis as shaped by social and political agents and institutions 1508–1886.* Unpublised Doctoral dissertation, European University Institute, Florence, Italy.

Kassam, K-A. (2014, March). *Wicked problems, diversity, and interdisciplinarity: The case of building anticipatory capacity for climate change.* Invited presentation, Department of Communication and Culture Colloquium, University of Calgary, Canada.

Kawagley, A.O., & Barnhardt, R. (1999). Education indigenous to place: Western science meets Native reality. In G.A. Smith and D.R. Williams (Eds.) *Ecological education in action: On weaving education, culture, and the environment* (pp. 117–140). New York: SUNY Press.

Kazina, D., & Swayze, N. (2009). Bridging the gap: Integrating Indigenous knowledge and science in a non-formal environmental learning program. *Green Teacher, 86,* 25–28.

Keown, M. (2008). "Our sea of islands of migration": Migration and métissage in contemporary Polynesian writing. *International Journal of Francophone Studies, 11*(4), 503–522.

Kienetz, A. (1983). The rise and decline of hybrid (metis) societies on the frontier of western Canada and southern Africa. *The Canadian Journal of Native Studies, III*(1), 3–21.

Kimmerer, R.W. (2013). The fortress, the river and the garden: A new metaphor for cultivating mutualistic relationship between scientific and traditional ecological knowledge. In A. Kulnieks, D.R. Longboat, & K. Young (Eds.) *Contemporary Studies in Environmental and Indigenous Pedagogies* (pp. 49–76). Rotterdam: Sense.

Kimmett, C. (2009, 24 July). First Nation takes lead on solar power. *The Tyee.ca.* Retrieved July 1st, 2011 from http://thetyee.ca/News/2009/07/24/FirstNationSolarPower/.

Kincheloe, J., & Steinberg, S. (2008). Indigenous knowledges in education: Complexities, dangers, and profound benefits. In N.K. Denzin, Y.S. Lincoln, & L.T. Smith (Eds.) *Handbook of Critical Indigenous Research,* (pp.135–156). Thousand Oaks, CA: Sage.

King, T. (2003). *The truth about stories.* Toronto, ON: House of Anansi Press.

Kino-nda-niimi Collective (Eds.) (2014). *The winter we danced: Voices from the past, the future, and the Idle No More movement.* Winnipeg, MB: ARP Books.

Klassen, N. (2006). Can we still speak Chinook? *The Tyee.* Retrieved December 4th, 2008 from http://thetyee.ca/Life/2006/01/10/StillSpeakChinook/.

Korteweg, L., & Oakley, J. (2014). Eco-heroes out of place and relations: Decolonizing the narratives of *Into the Wild* and *Grizzly Man* through Land education. *Environmental Education Research, 20*(1), 131–143.

Korteweg, L., & Russell, C. (2012). Editorial: Decolonizing + Indigenizing = moving environmental education towards reconciliation. *Canadian Journal of Environmental Education, 17*, 5–14.

Kovach, M. (2010). *Indigenous methodologies*. Toronto: University of Toronto Press.

LaDuke, W. (2002). *The Winona LaDuke reader: A collection of essential writings*. Penticton, BC: Theytus Books.

Lawrence-Lightfoot, S. (2005). Reflections on portraiture: A dialogue between art and science. *Qualitative Inquiry, 11*(3), 3–15.

Lefèvre, K. (1989). *Métisse blanche*. Paris: Bernard Barrault.

Lemesianou, C.A., & Grinberg, J. (2006). Criticality in education. In K. Tobin & J. Kincheloe (Eds.), *Doing educational research* (pp. 211–233). Rotterdam, The Netherlands: Sense Publishers.

Lertzman, D. (2002). Rediscovering rites of passage: Education, transformation, and the transition to sustainability. *Ecology and Society*, 5(2): Article 30. Retrieved February 27th, 2007 from http://www.ecologyandsociety.org/vol5/iss2/art30/.

Lichtman, M. (2012). *Qualitative research in education: A user's guide (2nd Ed.)*. Thousands Oaks, CA: Sage.

Lickers, M. (2006). *Urban Aboriginal leadership*. Unpublished master's thesis, Royal Roads University, Victoria, BC.

Little Bear, L. (2000a). Jagged worldviews colliding. In Battiste, M. (Ed.), *Reclaiming Indigenous Voice and Vision*, (pp. 77–85). Vancouver, BC: UBC Press.

Little Bear, L. (2000b). Foreword. In Cajete, G. *Native science: Natural laws of interdependence*. Santa Fe, NM: Clear Light Publishers.

London, L., Nell, V., Thompson, M.L., & Myers, J.E. (1998). Health status among workers in the Western Cape—Collateral evidence from a study of occupational hazards. *South African Medical Journal*, 88(9), 105–110.

Lotz-Sisitka, H. (2002). Weaving cloths: Research design in contexts of transformation. *Canadian Journal of Environmental Education*, 7(2), 101–124.

Loughland, T., Reid, A., Walker, K., & Petocz, P. (2003). Factors influencing young people's conceptions of environment. *Environmental Education Research*, 9(1), 3–20.

Lowan-Trudeau, G. (2015). Teaching the tension: Indigenous land rights, activism, and education in Canada. *Education Canada*, 55(1), 44–47.

Lowan-Trudeau, G. (2014). Considering ecological métissage: To blend or not to blend? *Journal of Experiential Education*, 37(4), 351–366 [originally published online first Dec. 18, 2013].

Lowan-Trudeau, G. (2013a). Indigenous environmental education research in North America: A brief review. In R. Stevenson, M. Brody, J. Dillon, & A. Wals (Eds.), *International handbook of environmental education research*. New York: Routledge.

Lowan-Trudeau, G. (2013b). Idle No More: What is it and why (should) it matter to environmental educators? *EECOM News*. Retrieved September 22nd, 2014 from http://www.eecom.org/index.php?option=com_content&view=category&id=15: member-news&layout=blog&Itemid=28&lang=en.

Lowan-Trudeau, G. (2012b). Methodological métissage: An interpretive Indigenous approach to environmental education research. *Canadian Journal of Environmental Education, 17*, 113–130.

Lowan, G. (2012a). Expanding the conversation: Further explorations into Indigenous environmental science education theory, research, and practice. *Cultural Studies in Science Education, 7*, 71–81.

Lowan, G. (2011a). Ecological métissage: Exploring the third space in Canadian outdoor and environmental education. *Pathways: The Ontario Journal of Outdoor Education, 23*(2), 10–15.

Lowan, G. (2011b). *Navigating the wilderness between us: Exploring ecological Métissage as an emerging vision for environmental education in Canada*. Unpublished doctoral dissertation, University of Calgary, Canada.

Lowan, G. (2010, September). *Navigating the wilderness between us: Ecological métissage as an emerging vision for environmental education in Canada*. Keynote address presented at the Council of Outdoor Educators of Ontario Conference, Camp Kinark, ON.

Lowan, G. (2009). Exploring place from an Aboriginal perspective: Considerations for outdoor and environmental education. *Canadian Journal of Environmental Education, 14*, 42–58.

Lowan, G. (2008). Paddling tandem: A collaborative exploration of Outward Bound Canada's Giwaykiwin Program for Aboriginal youth. *Pathways: The Ontario Journal of Outdoor Education, 20*(1), 24–28.

Lowan, G. (2007). Outward Bound Giwaykiwin: Connecting to land and culture through Indigenous outdoor education. *Pathways: The Ontario Journal of Outdoor Education, 19*(3), 4–6.

MacDougall, B. (2006). Wahkootowin: Family and cultural identity in northwestern Saskatchewan Metis communities. *The Canadian Historical Review, 87*(3), 431–462.

MacGregor, R. (2003). *Escape: In search of the natural soul of Canada*. Toronto: McLelland and Stewart.

Mack, E., Augare, H., Cloud-Jones, L.D., David, D., Gaddie, H.Q., Honey, R.E., Kawagley, O., Plume-Weatherwax, M.L., Fight, L.L., Meier, G., Pete, T., Leaf, J.R., Returns From Scout, E., Sachatello-Sawyer, B., Shibata, H., Valdez, S., & Wippert, R. (2012). Effective practices for creating transformative informal science education programs grounded in native ways of knowing. *Cultural Studies of Science Education, 7*(1), 49–70.

Macy, J. (2007). *World as lover, world as self: Courage for global justice and ecological renewal*. Berkeley, CA: Parallax.

Malenfant, E.C., Lebel, A., & Martel, L. (2010). *Projections of the diversity of the Canadian population*. Ottawa, ON: Statistics Canada.

Mallet, V.N. (2010). *Les Métis Acadiens de la Baie des Chaleurs*. Shediac Cape, NB: Victorin N. Mallet.

Martusewicz, R. A. (2013). Toward an anti-centric ecological culture: Bringing a critical ecofeminist analysis to ecojustice education. In A. Kulnieks, K. Young, & D. Longboat (Eds.) *Contemporary studies in environmental and Indigenous pedagogies: A curricula of stories and place* (pp. 225–240). Rotterdam: Sense Publishers.

Mateata-Allain, K. (2008). Orality and Maohi culture. *Shima: The International Journal of Research into Island Cultures*, *3*(2), 1–9.

McDermot, G. (1993). *Raven: A trickster tale from the Pacific Northwest*. New York: Scholastic.

McGinnis, M.V. (1999). A rehearsal to bioregionalism. In M.V. McGinnis (Ed.) *Bioregionalism* (pp. 1–9). New York: Routledge.

McKeon, M. (2012). Two-Eyed Seeing into environmental education: Revealing its "natural" readiness to Indigenize. *Canadian Journal of Environmental Education*, *17*, 131–147.

Menzies, C.R. (Ed.) (2006). *Traditional ecological knowledge and natural resource management*. Lincoln, NE: University of Nebraska Press.

Merchant, C. (2004). *Reinventing Eden: The fate of Nature in Western culture*. New York: Routledge.

Métis National Council. (2002). *Who are the Métis?* Retrieved April 10th, 2010 from http://www.metisnation.ca/who/index.html.

Milan, A., Maheux, H., & Chui, T. (2010). *A portrait of couples in mixed unions*. Ottawa, ON: Statistics Canada.

Miller, B.G. (2011). *Oral history on trial*. Vancouver, BC: UBC Press.

Myers, D. (2009). Two-eyed seeing in a school district. *Green Teacher*, 86, 39–40.

Naess, A., & Rothenberg, D. (1990). *Ecology, community and lifestyle*. Cambridge: Cambridge University Press.

Narine, S. (2009). Saddle Lake leads way with environmental water treatment. *Alberta Sweetgrass*, *16*(6), 1.

Natural Resources Canada (2014). *Government of Canada accepts recommendation to impose 209 conditions on Northern Gateway proposal*. Ottawa, ON: Natural Resources Canada.

Newberry, L. (2012). Canoe pedagogy and colonial history: Exploring contested space of outdoor environmental education. *Canadian Journal of Environmental Education*, *17*, 30–45.

Nguyen, N.H.C. (2005). Eurasian/Amerasian perspectives: Kim Lefevre's *Métisse Blanche* and Kien Nguyen's *The Unwanted*. *Asian Studies Review*, *29*, 107–122.

Niort, J.F. (2007). *Du code noir au code civil: jalons pour l'histoire du droit en Guadaloupe: Perspectives compares avec la Martinique, la Guyane et la République d'Haiti*. Paris: L'Harmattan.

Nute, G.L. (1987). *The voyageur*. St. Paul, MN: Minnesota Historical Society Press.

Oguri, Y. (2010, March). *Challenges in Japan to build community livelihood in a more sustainable way*. Invited presentation to the University of Calgary Faculty of Education, AB.

Owens, L. (2001). As if an Indian were really an Indian: Native American voices and postcolonial theory. In G.M. Bataille (Ed.), *Native American representations: First encounters, distorted images, and literary appropriations* (pp. 11–24). Lincoln, NE: U of Nebraska Press.

Ozog, S. (2012). *Towards First Nations energy self-sufficiency: Analyzing the renewable energy partnership between Tsou-ke Nation and Skidegate Band*. Unpublished master's thesis, University of Northern British Columbia, Prince George, BC.

Palmer, J.A., Suggate, J., Robottom, I., & Hart, P. (1999). Significant life experiences and influences in the development of adults' environmental awareness in the UK, Australia, and Canada. *Environmental Education Research*, *5*(2), 181–200.

Pashagumskum, S. (2014). First Nations student engagement in secondary school: Enhancing student success in a Northern Eeyou community. Unpublished PhD dissertation, Lakehead University, Thunder Bay, Ontario.

Payne, P. (1999). The significance of experience in SLE research. *Environmental Education Research*, *5*(4), 365–381.

Pepper, F.C., & White, W.A. (1996). *First Nations traditional values*. Victoria, BC: Aboriginal Liaison Office, University of Victoria.

Pieterse, J.N. (2001). Hybridity, so what? The anti-hybridity backlash and the riddles of recognition. *Theory, Culture & Society*, *18*(2–3), 219–245.

Pieterse, J.N. (1996). Globalisation and culture: Three paradigms. *Economic and Political Weekly*, *31*(23), 1389–1393.

Podruchny, C. (2006). *Making the voyageur world: Travelers and traders in the North American fur trade*. Toronto, ON: University of Toronto Press.

Portman, T.A.A., & Garrett, M.T. (2006). Native American healing traditions. *International Journal of Disability, Development and Education*, *53*(4), 453–469.

Raffan, J. (2008). *Emperor of the North: Sir George Simpson and the remarkable story of the Hudson's Bay Company*. Toronto, ON: Harper Collins Publishers Ltd.

Rahier, (2003). Métis, Mulatre, Mulato, Mulatto, Negro, Moreno, Mundele, Kaki, Black…In P.C. Hintzen & J.M. Rahier (Eds.), Problematizing Blackness: Self-ethnographies by Black immigrants to the United States (pp. 85–112). New York: Routledge.

Ramsay, R. (2008). In the belly of the canoe with Ihimaera, Hulme, and Gorodé. The waka as a locus of hybridity. *The International Journal of Francophone Studies*, *11*(4), 559–579.

Reid, P., & Bringhurst, R. (1996). *The raven steals the light*. Vancouver, BC: Douglas & McIntyre.

Reid, A., Teamey, K., & Dillon, J. (2002). Traditional ecological knowledge for learning with sustainability in mind. *Trumpeter*, *18*(1), 113–136.

Richardson, C. L. (2004). *Becoming Metis: The relationship between the sense of Metis self and cultural stories*. Unpublished doctoral dissertation, University of Victoria, Victoria, BC.

Rivard, E. (2008). Colonial cartography of Canadian margins: Encounters and the idea of *métissage*. *Cartographica*, *43*(1), 45–66.

Root, E. (2010). This land is our land? This land is your land: The decolonizing journeys of White outdoor environmental educators. *Canadian Journal of Environmental Education*, *15*, 103–119.

Root, E., & Dannenman, K. (2009, September). *Kitakiiminan: Learning to live well together on traditional lands*. Paper presented at and published in the proceedings of Henrik Ibsen: The Birth of "Friluftsliv" A 150 Year International Dialogue Conference Jubilee Celebration North Troendelag University College, Levanger, Norway Mountains of Norwegian/Swedish Border.

Rosen, N. (2003). Demonstrative position in Michif. *Canadian Journal of Linguistics/Revue canadienne de linguistique* 48(1/2), 39–69.

Rosen, N. (2008). French-Algonquian interaction in Canada: A Michif case study. *Clinical Linguistics & Phonetics*, *22*(8), 610–624.

Rosser, E. (2006). Ambiguity and the academic: The dangerous attraction of pan-Indian legal analysis. *Harvard Law Review*, *119*, 141–147.

Roth, W.M. (2008). Bricolage, métissage, hybridity, heterogeneity, diaspora: Concepts for thinking science education in the 21st century. *Cultural Studies in Science Education*, *3*, 891–916.

Russell, C., & Fawcett, L. (2013). Moving margins in environmental education. In R. Stevenson, M. Brody, J. Dillon, & A. Wals (Eds.), *International handbook of research on environmental education* (pp. 360–374). New York: Routledge.

Saul, J.R. (2008). *A fair country: Telling truths about Canada*. Toronto, ON: The Penguin Group.

Scully, A. (2012). Decolonization, reinhabitation and reconciliation: Aboriginal and place-based education. *Canadian Journal of Environmental Education*, *17*, 148–158.

Sharpe, E., & Breunig, M. (2009). Sustaining environmental pedagogy in times of educational conservatism: A case study of integrated curriculum programs. *Environmental Education Research*, *15(3)*, 299–313.

Silverstein, M. (1997). Encountering language and languages of encounter in North American ethnohistory. *Journal of Linguistic Anthropology*, 6(2), 126–144.

Simpson, L. (2010). First words. In L. Davis (Ed.), *Alliances: Re/Envisioning Indigenous-non-Indigenous relationships* (pp. xiii–xiv). Toronto: University of Toronto Press.

Simpson, L. (2004). Anticolonial strategies for recovery and maintenance of Indigenous knowledge. *American Indian Quarterly*, *28*(3 & 4), 373–384.

Simpson, L. (2002). Indigenous environmental education for cultural survival. *Canadian Journal of Environmental Education*. (1), 13–35.

Smith, L.T. (1999). *Decolonizing methodologies: Research and indigenous peoples*. London and New York: Zed Books; Dunedin, New Zealand: University of Otago Press.

Snively, G. (2009). Money from the sea: A cross-cultural Indigenous science activity. *Green Teacher*, 86, 33–38.

Snively, G., & Corsiglia, J. (2000). Discovering Indigenous science: Implications for science education. *Science Education*, 85(1), 6–34.

Snow, J. (2005). *These mountains are our sacred places*. Calgary, AB: Fifth House. (Original work published 1977).

Snyder, G. (2003). *The practice of the wild*. Berkeley, CA: Counterpoint.

Statistics Canada (2008). *Canada Year Book*. Ottawa, ON: Statistics Canada.

Steele, A. (2011). Beyond contradiction: Exploring the work of secondary science teachers as they embed environmental education in curricula. *International Journal of Environmental & Science Education*, 6(1), 1–22.

Steinberg, S. (2006). Critical cultural studies research: Bricolage in action. In K. Tobin & J. Kincheloe (Eds.), *Doing educational research*, (pp. 117–137). Rotterdam, The Netherlands: Sense Publishers.

Stephens, S. (2000). *Handbook for culturally responsive science curriculum*. Fairbanks, AK: Alaska Native Knowledge Network.

Stibbe, A. (2004). Environmental education across cultures: Beyond the discourse of shallow environmentalism. *Language and Intercultural Communication*, 4(4), 242–260.

Stoler, A. (1992). Sexual affronts and racial frontiers: European identities and the cultural politics of exclusion in colonial Southeast Asia. *Comparative Studies in Society and History*, 34(3), 514–551.

Sutherland, D., & Swayze, N. (2012). Including Indigenous knowledges and pedagogies in science-based environmental education programs. *Canadian Journal of Environmental Education*, *17*, 80–96.

Swayze, N. (2011). *Engaging Indigenous urban youth in environmental learning: The importance of place revisited*. Unpublished master's thesis. University of Manitoba, Winnipeg, MB.

Swayze, N. (2009). Engaging Indigenous urban youth in environmental learning: The importance of place revisited. *Canadian Journal of Environmental Education*, *14*, 59–72.

Takano, T (2005). Connections with the land: Land-skills courses in Igloolik, Nunavut. *Ethnography*, 6(4), 463–486.

Thomashow, M. (1996). *Ecological identity*. Cambridge, MA: MIT Press.

Thompson, G. (2009). Progress through polling: Why we should conduct polling about environmental education. *EECOM News*, *5*, 1–3.

Troupe, C.L. (2009). *Métis women: Social structure, urbanization and political activism, 1885–1980*. Unpublished master's thesis, University of Saskatchewan, Saskatoon, SK.

Tobin, K. (2006). Qualitative research in classrooms: Pushing the boundaries of theory and methodology. In K. Tobin & J. Kincheloe (Eds.), *Doing Educational Research*, (pp. 3–14). Rotterdam, The Netherlands: Sense Publishers.

Tsilhqot'in Nation v. British Columbia, SCC 44 (2014).

Tuck, E., McKenzie, M., & McCoy, K. (2014). Land education: Indigenous, post-colonial, and decolonizing perspectives on place and environmental education research. *Environmental Education Research*, *20*(1), 1–23.

Turner, N. (2005). *The Earth's blanket: Traditional teachings for sustainable living*. Vancouver, BC: Douglas & McIntyre.

Vansina, J. (1961/ 1996). Oral tradition and historical methodology. In D.K. Dunaway, & W.K. Baum (Eds.), *Oral history: An interdisciplinary anthology, 2nd edition* (pp. 121–125). Lanham, MD: Altamira Press. (Original work published 1961).

Vink, M. J., A. Dewulf, and C. Termeer. 2013. The role of knowledge and power in climate change adaptation governance: A systematic literature review. *Ecology and Society 18*(4): 46. Available at http://dx.doi.org/10.5751/ES-05897–180446.

Vizina, Y. (2008). Métis culture. In C. Avery and D. Fichter (Eds.), *Our legacy: Essays* (pp. 169–182). Saskatoon, SK: University of Saskatchewan.

Waldram, J.B. (2000). The problem of "culture" and the counseling of Aboriginal peoples. In Brass, G.M., Kirmayer, L.J., & Macdonald, M.E. (Eds.), *Report no. 0: The mental health of Indigenous peoples: Proceedings of the Advanced Study Institute McGill Summer Program in Social & Cultural Psychiatry and the Aboriginal Mental Health Research Team* (pp. 145–158). Montréal, QC: Culture and Mental Health Research Unit.

Welsch, W. (1999). Transculturality: The puzzling form of cultures today. In M. Featherstone & S. Lash (Eds.) *Spaces of culture: City, nation, world*, pp. 194–213. London: Sage.

Wildcat, M., McDonald, M., Irlbacher-Fox, S., & Coulthard, G. (2014). Learning from the land: Indigenous land-based pedagogy and decolonization. *Decolonization: Indigeneity, Education & Society*, 3(3), i–xv.

Willinsky, J. (2006). When the research is over, don't turn out the lights. In K. Tobin & J. Kincheloe (Eds.), *Doing educational research*, (pp. 439–456). Rotterdam, The Netherlands: Sense Publishers.

Wilson, N. (2008). A waka ama journey: Reflections on outrigger canoe paddling as a medium for epistemological adventuring. *Pathways: The Ontario Journal of Outdoor Education*, *21*(1), 19–23.

Wilson, S. (2008). *Research is ceremony: Indigenous research methods*. Halifax, NS: Fernwood Publishers.

Zembylas, M., & Avraamidou, L. (2008). Postcolonial findings of space and identity in science education: Limits, transformations, prospects. *Cultural Studies in Science Education*, *3*, 977–998.

INDEX

A

Aboriginal 1, 2, 3, 7, 9, 11, 14, 22, 24–26, 32–39, 41–53, 55, 60, 62, 69, 75, 77–80, 82, 83, 85, 86, 88–95, 97, 98, 104, 111–113, 116–121
Aboriginal ancestry 37– 39
Aboriginal communities 1, 2, 33, 104
Aboriginal cultures 14, 42, 45, 46, 51, 88, 89, 98, 112, 120
Aboriginal education 49, 98
Aboriginal languages 9, 48, 51, 52
Aboriginal mythologies 33
Aboriginal participants 11, 89, 91
Aboriginal scholars 45, 120
Aboriginal spirituality 43, 44, 50, 51
Acadia 34
Acadian 18, 34, 82, 88, 89
Accountability 21, 27, 105
Achebe, C. 105
Activism 108, 127, 128
Activist 66, 74
Adaptability 47
Adopting 29, 31, 49, 50, 52, 91, 109
Advocate 5, 6, 73, 109
Advocated 65, 98, 130
Advocates 28, 64
Advocating 67, 130
Africa 40, 41
Agriculture 38, 43, 47, 61
Agyeman, J. 102, 128
Aha moment 23, 82, 104
Aikenhead, G. 102
Alaska 76
Alaska Rural Systemic Initiative 76
Alberta 1, 7, 11–13, 36, 60, 82, 97, 127
Algonquian 6
Alliances 49, 110, 127
Allies 28, 127
America 7, 16, 25, 27, 35, 36, 39, 40, 42, 50, 56, 60, 66, 76, 79, 98, 120
American 12, 14, 45, 46, 55, 60, 67, 73, 83
American Indian 73

American indigenous 60
American pacific 83
American-style 46
Ancestors 7, 11, 32, 34, 39, 82
Ancestral 7, 43, 48, 76, 121
Ancestry 6, 7, 33, 36–41, 48, 55, 82, 84, 89, 90, 129
Anglican 43
Anglo-Canadian 11, 104
Anglo-Protestant 51
Animals 14, 69, 74, 78, 97, 105, 108, 116, 117
Animate 90
Anishnaabe 54, 61, 66, 74, 94
Anthropocentric 16
Anthropology 69
Appropriation 11, 80, 89
Arctic 109, 116, 126
Art 75, 92, 97
Arts, The 92, 114
Asia 40, 52
Asian 40, 52
Assimilation 69, 96
Authenticity 59

B

Baie des Chaleurs 34
Bakker, P. 48
Barnhardt, R. 9, 10, 19, 65, 68, 71, 76, 94, 96, 129
Basque 34
Bastien, B. 9, 22, 25, 26
Batoche 33
Battiste, M. 9, 70, 95
Beaver 10, 120
Belgian 40
Beluga 109
Berry, K. 12, 17, 18, 20, 24, 26
Berry, W. 12, 45
Bhabha, H. 20, 41
Bible 106
Bicultural 48, 56
Biodiversity 66, 67
Biographical 23, 25, 26, 103
Bioregionalism 59, 61, 71, 72, 74, 90, 98, 115, 120, 129
  Aberley, D. 72, 74
  Dodge, J. 59, 61, 71, 72
Birch 11
Birds 77, 82, 83
Birth 34, 44
Blackfoot 22, 67
  Siksikaitsitapi 22
Blend 18, 40, 43, 44, 51, 54, 55, 57, 59, 61, 63, 65, 66, 67, 69, 71, 73, 75, 77, 79, 94–96, 122
  blended 6, 15, 33, 43, 47, 48, 81, 86, 94–96
  blended, approach 95
  blended, hybrid 94
  blending 5, 29, 31, 41, 43, 48, 55, 56, 65, 71, 86, 94, 95–98
Borders 27, 128, 129
Boreal forest 7, 59, 79
Bow River 7
Braid 19
Bricolage 17, 18, 20, 29, 56, 63, 71, 86, 94, 95, 98, 115, 130
  birdhouses 18
  bricoleur 18, 19
  cobbling 18
Bride 44
Bridge 5, 96, 97
Bridging the Gap 75, 77
British 1, 2, 7, 8, 9, 34, 36, 52, 82, 83, 88, 104, 108, 126, 127
British Columbia 2, 36, 52, 83, 126, 127

C

Cajete, G. 16, 26, 28, 38, 45, 51, 52, 55, 60, 70–73, 78, 87, 101, 102, 120
Calgary 7
Campbell, M. 40

Canadian 5, 6, 8–12, 14, 15, 22, 26, 32, 34, 40, 44, 46, 49, 80, 82, 84, 88, 89, 92, 97, 101–103, 107, 110, 111, 114, 117–119, 121, 122, 128
Canadian Rocky Mountains 82
Canadians 3, 8, 9, 12, 13, 49, 53, 107, 120, 121
Canoe 8, 10, 11, 40, 60, 79, 84, 112, 119–122, 133
canoeing 7, 12, 121, 122
Cape Breton 6, 69, 77, 88, 102
Bartlett, C. 6, 48, 75, 77, 102, 105
Hatcher, A. 6, 48, 69, 75, 77, 102, 105
Integrative Science 6, 69, 75, 77, 83, 93, 96, 102, 111
Marshall, A. & M. 6, 48, 75, 102, 105
Two-Eyed Seeing 6, 48, 69, 75, 77, 102, 105
Caribbean 41
CBC 2, 62
Cedar 11, 50, 85, 123
Cedar sprig 50
Ceinture fléchées 51
Ceremony 27, 44, 54, 62, 70, 74, 107, 116, 123
Chambers, C. 5, 18, 29, 42, 114
Cherokee-American 7
Children 35, 40, 47, 52, 86, 97
Chinese 52
Chinook Wawa/Jargon 49, 52, 53, 64
Christian 14, 44, 50, 105
Christianity 106
Circle 66, 71, 73, 102
Clandinin, D. J., & Connelly, F. M. 23
Climate Action Network 10
Coast 7, 55, 78, 83, 85, 88, 116, 127
Cohen Commission 126
Collaboration 6, 53, 110, 119, 129, 131
Collaborative 25, 126
Colonial 19, 40, 41, 50, 52, 53, 67, 110, 119
colonial mentality 53, 119
colonial processes 110
colonial project 50, 53, 67
colonialism 9, 20, 60, 67, 128, 130
colonization 9, 33, 53, 89
colonizer 53
Community 1, 2, 6, 21, 27, 31, 36, 44, 45, 46, 47, 52, 59, 61, 62, 70, 76, 79, 83, 85, 88, 103–106, 110, 113, 116, 120, 131
community-based 29, 106
community-connected 115
community-developed 76
Congo 40, 41
Conservation 12, 67, 109
Conservative Party 13
Consultation 27, 76
Coulthard, G. 111
Cree 20, 37, 48
Creolization 56
Creolized 63
Creswell, J. W. 19, 104
Cross-cultural 35, 75, 78, 107, 118
Cultural 5, 6, 8, 9, 10, 11, 14, 15, 17, 18, 20, 22, 23, 24, 27, 28, 31–35, 38, 39, 41, 47, 48, 49, 51–53, 55, 57–65, 67, 68, 71, 73, 74, 76, 77, 80, 83–88, 90, 91, 93, 96, 102, 104, 105, 110, 111, 113, 114, 116, 117, 119, 121, 122, 127, 129, 130, 133
cultural identity 64, 90
cultural integration 71
culturally based 6, 115
culturally complex 15, 101, 103, 107
culturally diverse 52
culturally plural 57
culturally rooted 2, 29, 102, 125, 126, 130
Curriculum 65, 92, 97
Customs 21, 39, 44, 50
Cuthbertson, B. 59
Czikszentmihalyi, M. 11

D

Dancer 9
Dean, M. 11, 120
Decolonization 53, 66, 67, 75, 84

Deep ecologist 73, 74
Drengson, A. 73, 74
Macy, J. 73
Naess, A. 26, 45, 71, 73, 114
Rothenberg, D. 26, 45, 71, 73, 114
Delgamuukw decision 22
Denzin, N. 17, 23, 82
Diaspora 56, 63, 75
Diasporic 63, 64, 131
Dichotomy 129
Dillon, J. 67, 86, 87, 129
Dissertation 17, 34, 41, 82
Dogs 90, 108, 116, 117
dog sledding 90
Donald, D. 5, 18, 19, 29, 72
Dreams 11, 70, 93
Dualistic 65
Dumont, G. 32

E

Earth 3, 45, 54, 56, 90
earth-based 88
East 7, 34, 47, 83
East Asia 40
East Coast 7, 83
Eastern 7, 11, 82
eastern foothills 82
eastern woodlands 11, 25
Ecofeminism 128
Ecological 5, 6, 8, 10, 12, 13, 15, 17, 21, 23, 26, 31–33, 41, 49, 55, 56, 59, 61, 64–66, 70–74, 77, 79–82, 85–88, 90, 93–95, 99, 101–103, 105, 108, 111, 113, 114, 117, 119, 120–122, 125, 126, 128–131
ecological diversity 73, 120
ecological identities 15, 17, 23, 81, 85–87, 90, 93, 120, 121
ecological identity 5, 10, 12, 15, 41, 64, 80, 86–88, 90, 111, 117, 119, 120–122
ecological knowledge 5, 6, 15, 23, 56, 65, 66, 94, 99, 102, 103, 105, 108, 126, 128
ecological métissage 5, 8, 13, 15, 32, 49, 71, 72, 81, 82, 86, 93, 111, 125, 130
ecological philosophies 55
Ecology 26, 45, 59, 71, 73, 74, 98, 101, 129
Economic 1, 12, 13, 22, 28, 59, 70, 119
Eco-pedagogical 91
Eco-philosopher/Ecosopher 12, 87
Ecosophy 74, 130
Ecosystem 2, 66, 78, 88, 97, 109, 126, 127
Edge, L. 32, 37, 43–48
Education 2, 5, 7, 15, 22–28, 32, 33, 43, 47–49, 53, 56, 59, 65, 71, 75–80, 82–86, 90, 93, 95–98, 101, 102, 105, 107, 108, 111, 113, 115–117, 119, 121, 123, 125–129, 133
Educational 1, 3, 7, 15, 47, 53, 57, 62, 63, 69, 70, 75, 76, 93, 102, 103, 113, 128, 133
educational contexts 15, 102, 103
educational experiences 15, 76, 102
educational practices 57, 76, 93
educational programs 70
educational research 133
Educators 1, 5, 6, 15, 17, 18, 26, 60, 62, 63, 65, 69, 70, 71, 75, 77, 80–91, 93, 95, 97, 99, 102, 103, 106–109, 111, 113–115, 117, 121, 125, 127–131
Elder 1, 27, 44–48, 69, 75–77, 79, 85, 88, 89, 91, 93, 102, 109, 113, 116
Elementary school 7, 62, 105
England 31, 104, 107
English 7, 9, 42, 48, 50, 52, 80, 105, 108, 128, 129
English-language 64
English-speaking 11, 37
Environment 13, 26, 52, 88, 90, 92, 97, 108, 113, 117, 121, 122
Environmental 1, 2, 5, 6, 7, 13, 15, 17, 22–27, 32, 53, 56, 65, 66, 70, 71, 75, 77–91, 93, 95–99, 102, 103, 107, 108, 111, 113–115, 117, 119, 121, 123, 125–130
environmental education 2, 5, 15, 22–24, 26, 27, 32, 53, 56, 65, 75, 77–80,

82–86, 95–98, 102, 111, 113, 115, 117, 119, 121, 123, 126–129
environmental educators 5, 6, 15, 17, 70, 80, 81, 83–91, 93, 95, 97, 99, 103, 115, 121, 125, 130
environmental ethic 56, 127
environmental justice 128
environmental philosophers 26, 130
environmental problems 66, 113, 125
environmental research 25
environmentalism 58
environmentalists 12
Epiphanic 23, 82, 85, 92, 93, 104, 110, 115
Erasmus, P. 43
Esquimaux 37
Ethics 19, 27, 69, 92
Ethnicity 5, 28, 38, 57, 62, 89–91, 96, 117
Ethnobotany 55, 77
Eurasian 40, 97
Euro-American 45, 55
Euro-Canadian 11, 33, 34, 38, 41, 42, 44, 46, 55, 67, 83, 86–89, 91, 112, 113, 115
Euro-Indian 33
Euro-North American 45
Europe 47, 60, 101, 104, 107
European 3, 6, 7, 9, 14, 17, 19, 20, 32–37, 39, 40, 41, 43, 44, 46–48, 52, 53, 57, 60, 63, 84, 86, 88, 90, 103–105, 119, 130
European Explorers 35
Europeans 9, 38, 41, 42, 46, 53, 119

F

Families 7, 33–36, 38–41, 46, 47, 97, 107
Family 6, 7, 8, 11, 22, 25, 27, 34, 38, 39, 42, 44, 46, 47, 53, 64, 70, 79, 86, 88, 104, 110, 112, 119, 120, 122
Farm 2, 12, 45, 46, 52, 97, 110
farmer 12, 45, 61
farming 12, 79
Fashion 1, 44
Fawcett 128
Federal 1, 13, 35, 127
Federally recognized 35
Female 55, 109
Femininity 13
Feminism 58
Fire 37, 54, 79, 84
First-generation 102, 103
Fish 69, 107
fisheries 126
fishing 7, 12, 43, 45, 47, 61, 78, 79
Flow 11, 23
Folk 12, 43, 57
Folk Catholicism 43
Folklore 59
Forest 7, 12, 59, 79, 86, 87, 120, 122
Fossil 10, 13, 26, 61
Fox 14
Framework 15, 53, 56, 75, 78, 86
Francis, D. 10, 122
Francophone 9, 46
French-dené 39
French-speaking 41
Fundamentalism 58
Fur trade 11, 34, 35, 38, 44, 50, 51, 119, 120
bourgeois 50, 51
fur traders 35, 37, 38
highland boatmen 50

G

Gaffes 116
Garden 105
Genealogy 34, 35
Generations 7, 36, 62, 66, 74, 98
Genetic 34
Geographic 28, 59, 60
Geographical 24–28, 39, 52, 59, 61, 72, 87, 88, 91, 101, 102, 126
Geography 32, 33, 43, 90, 110, 119–121
German 57, 83
German-Canadian 63
Ghettoization 58
Ghost 79

Globalization 56, 60, 63, 126
Government 1, 2, 9, 11, 13, 38, 46, 50, 52, 64, 107, 126
Grandfather/s 34, 54
Grandmother/s 34, 54, 97, 105
Grant, J. 31
Graveline, F. J. 9, 14, 33, 113
Great Lakes 10, 11, 34, 121
Greek 13–16
Gruzinski, S. 35
Guadalupe 41
Guyana 79

H

Habitant 10, 11, 49
Haiti 41
Half-Aboriginal 32
Half-European 32
Haluza-Delay, R. 127, 128
Hart, P. 22, 23, 86
Hasebe-Ludt, E. 5, 18, 29
Hawaii 52, 79, 116
Hemingway, E. 39
Henderson, C. 2
Henley, T. 79
Herbs 106, 123
Heterogeneity 63
Heterogeneous 63
Hogue, M. 102
Holism 69, 96
Holistic 1, 19, 107
Holocaust 67
Home 10, 12, 16, 28, 47, 48, 52, 58–61, 63, 72, 83, 87, 88, 90, 101, 110, 114
  homeland 36, 62
  homes 2, 47, 63
  homesteads 49
Homogeneity 58, 57, 60, 62, 126
Honour 3, 27, 95, 98, 126, 127
Honouring 72, 73, 94, 118
Horsfall, G. A. 47
House 2, 11, 106, 121
Housing 46
Hudson Bay Company 92
Hunt 12, 32, 43–47, 61, 108, 109
Hybrid 61, 63, 65, 94
Hybridity 53, 56, 58, 63, 64, 65

I

Igloolik 76
  Paariaqtuqtut 76
Immigrants 12, 15, 97, 102, 104, 114, 122
Immigration 8, 101, 103, 114
Indian 32, 33, 43, 47, 73
Indians 35, 38
Indigene 59
Indigenize 50, 51, 66
Indigenous 2, 3, 5–11, 13, 15–29, 31, 39, 40, 43–45, 47, 50–56, 60, 65–84, 87, 89, 91–99, 101–103, 105–110, 112–116, 125–131
  Indigenous approaches 19, 26, 71, 94, 96, 129
  Indigenous communities 2, 9, 22, 55, 106, 108, 126
  Indigenous contexts 27, 80, 128
  Indigenous cultures 9, 19, 20, 22, 24, 26, 28, 55, 60, 65, 72, 91, 116, 129
  Indigenous ecological 6, 15, 17, 23, 65, 66, 74, 99, 102, 103, 105, 108, 128
  Indigenous education 71, 127
  Indigenous knowledge 6, 10, 20, 21, 28, 31, 54, 56, 66–71, 81, 89, 94–98, 102, 105–109, 113, 115, 126, 129, 130
  Indigenous languages 51, 72, 75
  Indigenous methodologies 20, 82
  Indigenous perspectives 83, 92, 106, 107, 130
  Indigenous philosophies 75, 91
  Indigenous research 21, 22, 24, 25, 26, 27, 103, 105
  Indigenous researchers 6, 20, 27
  Indigenous scholars 16, 22, 28, 56, 66, 67, 74, 87, 95

Indigenous students 65, 75, 76, 77, 83, 114
Indigenous territories 71, 72
Indigenous traditions 19, 20, 43, 73, 78, 79, 106, 129
Industrial 2, 56, 126
Industrial Revolution 56
Industry 1, 126
Infinity symbol 19, 20, 71, 78, 120, 122
Inner-city 77
Integrate 54, 66, 67, 75
integrated 2, 6, 69, 79, 86, 93, 95, 98, 105
integrating 56, 70, 76, 95, 128
integration 5, 6, 17, 29, 31, 56, 62, 70, 71, 96
Intercultural 2, 6, 8, 10, 12, 14–18, 20, 22, 24, 26, 28, 29, 32, 34, 36, 38, 40–42, 44, 46, 48, 50, 52–54, 56–58, 60, 62, 64, 66, 68, 70–72, 74, 76–86, 88, 90–96, 98, 102, 104, 106, 108, 110–122, 125–131, 133
intercultural bioregionalism 98
intercultural bricolage 98
intercultural collaboration 53, 110
intercultural conflicts 58, 115
intercultural cooperation 119, 127
intercultural engagement 56, 114
intercultural environmental 15, 17, 77, 80, 81, 82, 85, 86, 90, 93, 96, 111, 115, 117, 119, 125, 130
intercultural exchange 40, 108
intercultural mediators 14, 40
intercultural science 83, 98
Interculturalism 57, 58
Interdisciplinary 69, 75, 92, 97, 110, 114, 123
Inter-Indigenous 79, 131
Intermarriage 43, 53
Interpretation 12, 20, 35, 42, 43, 50, 59, 68, 103, 104
Interpretive 17–22, 24, 26–28, 77, 82, 103, 130
interpretive methodologies 20, 24
interpretive paradigms 18
interpretive research 24, 26, 130
interpretive scholars 22, 26
Interview 17, 23, 24, 25, 26, 27, 81–85, 98, 103, 104, 112, 117, 126, 128
Inuit 35, 45, 84, 109, 117
inukshuks 122
Inuktitut 76
Iroquois 38, 50
Island 1, 2, 25, 42, 88, 107
Isolated 12, 61, 64
Italian 61

J

Jackatar 34, 37
Japan 7, 79
Japanese 17, 52, 64, 76, 85, 89, 129
Jigs 31
Jokers 31
Justice 9, 28, 62, 68, 127, 128

K

Karahasan, D. 33–39, 41–47, 49–51
Kassam, K. A. 126
Kaupapa Maori 110
Kawagley, A. O. 9, 10, 19, 20, 65, 68, 71, 76, 94, 96, 129
Kazina, D. 75, 77
Kichigami/Lake Superior 61
Kimmerer, R. W. 6, 94
Kincheloe, J. 28, 68, 71, 133
Kino-nda-niimi Collective 110
Kovach, M. 17, 19–25, 27, 28, 82, 105

L

Labrador 34, 37, 39, 45, 46, 47
Laduke, W. 66, 74
Land-based 27, 47, 76, 107
Land-claim 22

Language 9, 19, 31, 37, 43, 48, 50–53, 58, 64, 65, 70, 72, 79, 80, 105, 106, 129
Latin 14, 36, 68, 80
Lawrence-Lightfoot, S. 23, 105
Legal 22
Lenape 7, 11
Lens 14, 33, 72, 75, 84, 108, 117, 130
Liminality 65
Linguistics 8, 32, 34, 37, 39, 40, 48, 51, 65, 104, 129
Local 13, 44, 50, 58, 59, 61, 62, 71, 76–79, 80, 85, 105, 113, 126
Loggers 12, 122
Louisiana 34
Lowan(-Trudeau), G. 5, 21, 42, 47, 75, 82, 102, 103, 106, 110, 115, 128
Lumberjack 10

M

Macgregor, R. 12
Malecite Métis 47
Manitoba 36, 46, 48
Maori 79, 110
Marital 41
Maritime provinces 36
Marriages 34, 35, 44
  morganatic marriage 44
Martusewicz, R. 128
Master's research 47, 75, 76, 93
Math 63, 64, 79, 97
Mccallum, T. 32, 37, 43, 44, 45, 46, 47, 48
Mcclachlin, Justice B. 127
Mediators 14, 40
Medicine Wheel 78
Membrane filter 2
Men 14, 35, 37, 40, 41, 44
Merchant, C. 12, 16
Mercury Poisoning 79
Mestizo 14
Metaphysical 16, 69, 79
Methodological 17–21, 23, 25, 27–29, 68, 72, 103, 130
  methodological bricoleur 19
  methodological development 103
  methodological influences 19
  methodological métissage 17, 19, 21, 23, 25, 27–29, 72
Methodologies 17, 20, 24, 29, 82, 105, 130
Methodologists 23
Methodology 17, 18, 19, 22, 29
Methods 26, 27, 66, 78
Metis 13–16, 38, 41, 45
  Détienne, M. 13
  dolos 14
  Vernant, J. P. 13
Métis 3, 6–9, 11, 14–20, 25, 26, 28, 29, 31–53, 61, 63, 71, 72, 77, 80, 83, 87, 89, 91, 98, 105, 117, 118, 119, 127–130
  Apihtaw'kosisan 37
  breeds 37
  buffalo 32
  buffalo-hunting 46
  burnt-wood/bois brûlé 37
  half-breed 36, 37
  half-burnt 37
  half-caste 37
  half-son 37
  Métis children 35, 47, 52
  Métis civilization 8
  Métis communities 31–35, 39, 41, 43, 45–49, 53
  Métis consciousness 34, 36
  Métis cultures 32, 38, 39, 43, 45, 46, 49
  Métis diversity 48
  Métis educators 77
  Métis elders 44–47
  Métis ethnogenesis 33
  Métis families 7, 34–36, 46, 47
  Métis flag 19, 71
  Métis geographies 32, 33
  Métis history 33, 35, 37, 41
  Métis homes 47
  Métis identity 29, 33, 35, 38, 39, 40
  Métis language 37
  Métis marriage 44
  Métis mentality 32, 41, 43, 80

Métis nation 8, 15, 36, 117, 118, 130
Métis peoples 14, 26, 32–37, 39, 40, 42, 48, 49, 72
Métis politics 36
Métis scholar 3, 8, 17, 20, 41, 43
Métis scrip 38
Métis settlement 47
Métis worldview 31–33, 35, 37, 39, 41, 43, 45–47, 49, 51, 53
Métis youth 47, 48
Michif 37, 48, 51, 52
Otipemisiwak 37
We-sahkotewenowak 37
Wissakodewinmi 37
Métissage 5, 7, 8, 13–15, 17–20, 21, 23, 25, 27–29, 31–35, 39, 40, 42–44, 49, 53, 56, 61, 63–65, 71, 72, 75, 81, 82, 86, 92– 95, 98, 103, 111, 115, 118, 119, 125, 128, 130
métisseur 29, 130
Minamata 79
Misappropriation 28, 97, 98
Miscegenation 14
Miscere 36
Missionary 43, 46
Mix 14, 17, 18, 20, 34, 40, 50, 61, 103
Mixed 6, 7, 9, 14, 33–38, 40–42, 48, 50, 52, 53, 55, 83, 84, 86–90, 93, 121
mixed ancestry 48, 90
mixed cultural 83, 86, 93, 121
mixed heritage 7, 38
mixed language 52
Mixing 5, 25, 51, 65, 95, 98
Mixtus 14, 36
Moccasins 51
Modern 1, 40, 57, 58, 66, 75, 120
Modernity 10, 13
Monocultural 57, 58, 59
Montana 36
Montreal 11, 47, 49, 50
Mother 7, 35, 44, 47, 54, 90
Mountains 7, 8, 10, 11, 72, 120, 127
Multicultural 9, 14, 15, 48, 56, 57, 102
Multiculturalism 56, 58, 63, 65
Multilogical 71
Multiple-Eyed Seeing 6
Mythologies 9, 33
Mythology 9, 14, 16

N

Nakoda Sioux 72
Nanabijou 61
Narrative 19, 21–25, 49, 50, 82, 92, 103, 105
narrative approach 19, 21
narrative inquiry 22, 23
narrative methodology 22
narrative portraits 105
narrative researchers 82
Narratives 19, 22, 23, 25, 53, 92, 103, 114
National 10, 13, 35, 39, 57, 58, 61, 121, 129
Nations 1, 2, 3, 8, 35, 39, 41, 54, 57, 58, 66, 67, 101, 103, 108, 118, 126, 128
Native 63, 66, 67, 70, 76, 97, 112, 113
Nativeness 39
Natural 5, 10, 11, 15, 16, 19, 25, 73, 77, 81, 87, 94, 101, 104, 120, 127
Nature 3, 8, 10, 11, 12, 13, 16, 19, 22, 32, 56, 68, 69, 86, 87, 89, 90, 94, 95, 96, 117, 121, 122, 123, 129, 130
Neocolonial 49
Nephew 89, 91
New Brunswick 34
New Zealand 79
Nigerian 105
Nomadic 46
Nomadism 38, 60
Non-Aboriginal 1, 2, 8, 25, 26, 55, 69, 89, 91, 92, 113, 118, 119
non-Aboriginal participants 89, 91
non-Aboriginal students 113
Non-Indigenous 3, 5, 6, 17, 20, 27, 29, 55, 56, 66, 67, 69, 70, 77, 80, 83, 91, 108, 110, 114, 126–128, 130, 131
non-Indigenous educators 69, 128
non-Indigenous participants 69

non-Indigenous researchers 6, 20
non-Indigenous students 77, 83, 114
Non-Métis 29, 32–34, 42, 49, 50, 52, 53
North 7, 10, 14, 16, 25, 27, 35, 36, 39, 40, 42, 46, 50, 55, 56, 60, 66, 73, 76, 79, 98, 107, 109, 120, 122
North Dakota 36
Northeastern 25, 34
Northern 7, 33, 34, 36, 59, 62, 85, 89, 104, 110, 122, 127
Northern Gateway Pipeline 127
Northwest 11, 34, 36, 39, 47, 49, 50, 52, 83, 86
Northwest Company 49
Northwestern 34, 49, 79, 85, 90
northwestern Ontario 79, 85, 90
northwestern States 34
Norwegian 7, 82, 88
Nunavut 76, 120

O

Ocean 31, 40, 104
Oceanography 92
Octopus 14
Oguri, Y. 79
Oil/Tar sands 127
Oji-cree 79, 80
Ontario 34, 36, 59, 60, 78, 79, 84, 85, 90, 104, 114, 120, 133
Ontology 64, 68, 98, 113
Oral 20, 22, 23, 34, 43, 77
Ottawa 50, 104
Outdoor 7, 59, 60, 75, 78, 79, 82, 84, 85, 88, 90, 133
outdoor education 7, 59, 90, 133
outdoor education sub-culture 59, 90
outdoor educator 59, 60, 90
Outward Bound Giwaykiwin 75
Owens, L. 7

P

Pacific 52, 78, 83, 86, 127
Pacific coast 78, 127
Pacific northwest 52, 83, 86
Pacific regions 52
Paddling 11, 133
Pakistani 17, 83, 86, 87, 129
Pan-Indianism 39, 60
Parents 7, 86, 122
Participants 6, 11, 15, 17, 18, 21–28, 69, 76, 81, 82, 86–89, 91, 94, 97–99, 102–106, 108, 109, 111, 112, 115–119, 122, 125, 127–130
Arthur 85, 88–91, 116, 118, 120, 121
Cedar Basket 88, 90, 92, 93, 97–99, 102, 118, 119, 122, 126
Kate 84, 85, 88–91, 95, 96, 118, 119, 121
Kathy 104, 106–108, 110, 122
Liatrus 83, 86, 87, 90, 91, 93–98, 118, 121
Natasha Little 82, 83, 88, 89, 91, 93, 94, 96, 98, 122
Olivia L'Hirondelle 83, 89, 90, 94–96, 97
Orange Blossom 83, 86, 92–94, 112, 113, 116
Skywalker 82, 86, 88, 91, 97, 99, 101, 102, 108, 112, 113, 115–117, 129
Sophia 104, 106, 107, 109, 110, 116, 126, 127
Thor 84, 88, 105, 117, 120, 121, 127, 130
Tong 84, 85, 87, 88, 114, 115, 129
Pedagogical 2, 6, 17, 29, 91, 92, 102, 108, 110, 113–115, 118, 125–128, 130
Pedagogies 95, 106
Pieterse, J. N. 5, 15, 53, 64, 65
Pilot study 6, 17, 102, 103, 105, 106, 108, 111, 112, 116, 122, 126, 127
Place-based 21, 29, 74, 129
Plants 7, 37, 45, 50, 69, 74, 77, 80, 108, 110
Plurality 57, 91
Podruchny, C. 35, 44, 46, 49, 50, 51
Polar bear 109
Policy 53, 66, 122

Political 8, 9, 13, 37, 39, 40, 67, 107, 110, 119, 122
Polynesia 40, 79
A'o/Ako 115, 116, 117
Wa'ka 40, 79
Portrait 23, 105
Positivist 19, 20
Post-secondary 8, 69, 79, 106
Potlach 2
Prairie 36, 46, 83, 84, 122
Preservationists 12
Problematized 21, 90
Protestant 43

Q

Qualitative 18, 20, 23, 24
Quantitative 19
Questions 15, 21, 23, 27, 32, 33, 41, 55, 56, 66, 67, 78, 81, 103, 111, 127

R

Race 11, 19, 65, 90, 91
Racial 64, 79
Racism 57, 61, 62, 64, 107
Raffan, J. 50
Rape 35
Reciprocity 21, 26, 27, 105, 112, 113, 115, 116
Reconciliation 131
Rediscovery 79
Reductionism 97
Reductionist 19, 96, 114
Reflexivity 8, 21, 24, 25
Refugees 12
Regional 8, 31, 35, 37, 39, 57, 59–62, 64, 88, 90, 91, 120
Relationship-building 113, 116
Relativism 68
Religion 43, 48, 63, 105
Research 8, 15, 17–28, 43, 47, 67, 68, 71, 80–82, 84, 86, 93, 102, 103, 105, 116, 125, 128, 129, 130, 133
researcher 1, 6, 18–22, 25–29, 71, 76, 82, 102, 104, 107, 116
Reserves 106, 113, 114
Resilience 73
Resistance 14, 33, 67
Respect 6, 18, 71, 73, 77, 91, 108, 112
Responsibility 45, 47, 62, 73
Revitalization 48, 97
Richardson, C. L. 20, 31, 33, 35–39, 41, 42
Rights 70, 73, 126
Rituals 44, 45, 50, 51, 74
Rivard, E. 32, 37
River 7, 11, 32, 33, 43, 46, 47, 49, 50, 79, 82, 87, 88, 112, 122, 126
River-lot 46
Rocky Mountains 7, 10, 11, 72
Roman Catholic 43
Romantic 11, 12, 109
Romanticism 59
Romanticizing 109
Root, E. 68, 70, 127
Roth, W. M. 18, 20, 29, 53, 57, 63, 64, 65, 71, 80, 98, 102, 113, 131
Rupert's Land 37, 46
Rural 7, 12, 76, 122
Russell, C. 111, 114, 128

S

Saddle Lake First Nation 1, 2, 5
Sage 123
Salmon 83, 122, 126
Salmon-bearing 127
Saskatchewan 7, 36, 46, 48
Saul, J. R. 8, 9, 14, 53, 57, 60, 101, 117–119, 121, 130
Sault St. Marie 34
Saulteaux 20
Sauna 88
Savage 34

Scandinavian 61, 88
Science 1, 3, 6, 15, 17, 22, 23, 26, 53, 56, 63–71, 75, 77–80, 83, 92–98, 101–103, 107, 109, 111, 114, 115, 125, 126, 128, 129, 133
scientia 68
scientific 56, 63, 65, 66, 109
scientists 18, 65–67, 126
Scottish 42, 46, 50, 61
Scully, A. 27, 102, 127
Sea 11, 45
Seashore 116
Self-identifying 39
Settlement 35, 36, 38, 43, 46, 47, 53, 119, 121
Settlers 37, 52, 101, 119, 120
Shona 105, 106, 129
Significant life experience (SLE) 86, 129
Simpson, L. 9, 28, 29, 45, 47, 51, 52, 56, 66–68, 70, 73, 75, 77, 91, 94, 97, 106, 115
Skiing 7, 12
Smith, L. T. 17, 19, 21
Smudging 116, 123
Snively, G. 29, 67, 68, 75, 78, 80, 102
Snowshoes 90
Snyder, G. 16, 87
Sociocultural 2, 26, 59, 62, 67, 98, 128
Socio-ecological 3, 112, 126
Socioeconomic 110, 122
Sociological 67
Solar 2, 5
Spiritual 31, 32, 43, 44, 50, 51, 52, 54, 66–68, 70, 71, 73, 96
Spring 11, 45
Spruce 11, 77
Squid 14
Steinberg 17, 18, 24, 25, 28, 68, 71
Sto:lo 17, 85, 88, 97, 98, 118, 129
Storytelling 77, 89
Student 63, 64, 70, 83, 85, 92, 107, 116, 118
Student-teacher 111
Swayze, N. 75, 77, 102
Sweatlodge 43, 44, 54, 116
Sweetgrass 123
Swiss 7, 83
Symbolism 51, 65, 70

T

Tamarack 77
Teaching 7, 8, 50, 69, 75, 83–85, 92, 93, 95–98, 113, 115, 116
Technocratic genocide 36
Technology 2, 3, 56, 78, 85
Teens 47, 85
Tension 25, 27, 33, 87, 110, 127
Thailand 79
Theoretical 18, 71, 72, 74, 128
Thomashow, M. 59, 86, 130
Thor 84, 88, 117, 120, 121
Thoreau, H. 130
Three-Eyed Seeing 101, 102, 103, 105–107, 109, 111, 125, 128, 130
Tobacco 27, 44, 50, 51, 77, 80
Traditional 1, 2, 7, 9, 19, 22, 27, 36, 40, 45, 47, 57–59, 65, 69, 70, 73, 77, 79, 80, 96, 105, 106, 108, 110, 120, 127
Traditional Ecological Knowledge (TEK) 56, 66–68, 93
Trans-disciplinarity 95
Transnational 63
Tribal 21, 60
Trickster 13
T'sou-ke First Nation 1, 2, 5
Chief Gordon Plumes 2

U

Uncomfortable 64, 115

V

Validity 67, 68
Vancouver Island 1, 2
Village 46, 79

Violence 12, 65
Vizina, Y. 36, 37, 43, 46, 48
Voyages 79
Voyageur 10, 11, 13, 32, 49–52, 59, 112
  calumet/pipe 44
  Canadien 37
  centennial voyageur pageant 11
  en dérouine 49
  hivernants 49
  Lachine 50
  Oiseau Rock 50
  voyageur ancestors 11
  voyageur baptism 50
  voyageur canoe 112
  voyageur rituals 50
  voyageur worldviews 49

W

Weaving 23, 69
Wedding 44
Welsch, W. 57, 58, 59, 60, 62, 63, 64, 91, 126
Western 1, 2, 3, 5–8, 10, 12, 13, 15–20, 23, 26, 28, 31, 32, 34, 37, 42, 44–48, 51, 53–56, 61, 64–72, 75–81, 83, 84, 91–99, 101, 102, 106, 109, 113–115, 125–127, 129, 130
  Western Canada 1, 7, 49, 127
  Western compartmentalism 69
  Western medicine 97
  Western science 6, 17, 19, 23, 26, 56, 65–69, 71, 77–79, 83, 93, 94, 96, 101, 109, 114, 115, 126, 129
White 3, 37, 38, 41, 65, 77, 84, 107, 127
Wilderness 12, 13, 16, 47, 84, 88
Wilson, N. 40, 79, 116, 117, 133
Wilson, S. 17, 22, 23, 27
Winnipeg 77
Women 35, 40, 41, 44, 49, 53, 82, 83, 84, 86–88, 91, 119
Woodlands 7, 11, 25
Worldview 9, 31–33, 35, 37, 39, 41, 43, 45–51, 53, 57, 60, 63, 66, 67, 73, 89, 97

Y

Youth 7, 26, 47, 48, 75–77, 83, 84, 99, 104, 106, 128

Z

Zimbabwe 104–106, 110
Zimbabwean 110
Zoo animals 97

(RE)THINKING ENVIRONMENTAL EDUCATION

GENERAL EDITORS: CONSTANCE RUSSELL & JUSTIN DILLON

The [Re]thinking Environmental Education book series is a response to the international recognition that environmental issues have taken center stage in political and social discourse. Resolution and/or re-evaluation of the many contemporary environmental issues will require a thoughtful, informed, and well-educated citizenry. Quality environmental education does not come easily; it must be grounded in mindful practice and research excellence. This series reflects the highest quality of contemporary scholarship and, as such, is positioned at the leading edge not only of the field of environmental education, but of education generally. There are many approaches to environmental education research and delivery, each grounded in particular contexts and epistemological, ontological and axiological positions, and this series reflects that diversity.

For additional information about this series or for the submission of manuscripts, please contact:

Constance Russell & Justin Dillon
c/o Peter Lang Publishing, Inc.
29 Broadway, 18th floor
New York, New York 10006

To order other books in this series, please contact our Customer Service Department:

(800) 770-LANG (within the U.S.)
(212) 647-7706 (outside the U.S.)
(212) 647-7707 FAX

Or browse by series:

WWW.PETERLANG.COM

www.ingramcontent.com/pod-product-compliance
Ingram Content Group UK Ltd.
Pitfield, Milton Keynes, MK11 3LW, UK
UKHW021839100926
8556IPUK00004B/440

9 781433 122361